The Cambridge Planetary Handbook

The *Cambridge Planetary Handbook* is an invaluable reference text, bringing together key facts and data on the planets and their satellites, discoverers and researchers. It summarizes many centuries' worth of data, from the earliest observations of the planets through to the most recent space probe findings. The author discusses the history, mythology and theories of the objects in our solar system, and provides a comprehensive information section with accurate and up-to-date data on the planets. The book contains excellent photographs and explanatory illustrations, along with numerous historical drawings from Galileo, Huygens, Herschel and other astronomers.

This book is a must for all astronomy enthusiasts, as well as academic researchers, students and teachers. Those unfamiliar with the sky will find this is a user-friendly guide written in clear, non-technical language.

Michael E. Bakich was born on 19 August 1953, in Steubenville, Ohio, USA. He has a BS in Astronomy from Ohio State University and an MA in Planetarium Education from Michigan State University. He currently lives in El Paso, Texas, and also leads tours to see eclipses and astroarchaeological sites. Michael has written numerous planetarium programs, gives lectures on planetary science to a wide range of audiences and is the author of the successful *Cambridge Guide to the Constellations*.

The Cambridge Planetary Handbook

MICHAEL E. BAKICH

CAMBRIDGE
UNIVERSITY PRESS

PUBLISHED BY THE PRESS SYNDICATE OF THE UNIVERSITY OF CAMBRIDGE
The Pitt Building, Trumpington Street, Cambridge, United Kingdom

CAMBRIDGE UNIVERSITY PRESS
The Edinburgh Building, Cambridge CB2 2RU, UK   www.cup.cam.ac.uk
40 West 20th Street, New York, NY 10011-4211, USA   www.cup.org
10 Stamford Road, Oakleigh, Melbourne 3166, Australia
Ruiz de Alarcón 13, 28014 Madrid, Spain

First published 2000

Printed in the United Kingdom at the University Press, Cambridge

Typeset in 10/12 Palatino [SE]

A catalogue record for this book is available from the British Library

Library of Congress Cataloguing in Publication data

Bakich, Michael E.
The Cambridge planetary handbook / Michael E. Bakich.
    p.   cm.
Includes bibliographical references and index.
ISBN 0 521 63280 3. – ISBN 0 521 63415 6 (pbk.)
1. Planets – Handbooks, manuals, etc.   I. Title.
QB601 B36   1999
523.4–dc21 99–10171   CIP

ISBN 0 521 63280 3 hardback

Contents

Preface

The flickering lights which play in the nighttime sky have been a source of unending fascination for tens of thousands of years. However early on it was observed that there were celestial lights which do not flicker. Indeed, as more observations were gathered, it was not simply the lack of scintillation which set these objects apart. They were among the brightest luminaries in the heavens. And they moved. They came to be called "planets," from a word which means "wanderer."

The stars also move, as can be proved by even casual observation over the course of a night. Their motion is regular. From east to west, night after night, year after year. Toward the north, stars seem to circle α Ursae Majoris – Polaris – Earth's North Star. This motion is also regular. Predictable. Seemingly unchanging.

It is against this regularity that the planets impose their motions. Mars, for example, may be in front of the stars of the constellation Aries one month, Taurus the next and Gemini the following month. Or it may linger in one constellation for several months. Jupiter, on average, takes a full year to migrate into the zodiacal constellation to its east. Saturn remains in front of individual star figures for even longer intervals. There are also nights which find the planets outside the traditional zodiac. Constellations such as Ophiuchus, Cetus and others lay claim to these wanderers for extended periods of time.

The general movement of the planets Mars, Jupiter and Saturn is eastward through the traditional zodiac of the ancients. From time to time, though, even this changes. The most striking facet of planetary observation is that these objects change their direction of motion.

Through the years, this motion has come to be understood. It is now known that the reversal of planetary motion (retrograde motion) occurs when the Earth, orbiting the Sun much faster than do the outer planets, passes each of them as they are near opposition. At these times, each of these bodies will appear to slow down, stop (a stationary point) and reverse direction, only to resume normal (direct) motion when the Earth has passed.

We live in an exciting time in the history of our solar system. Since antiquity, three additional planets have been discovered. Large telescopes have studied the cloud and surface features on all the planets. Their positions and motions have been calculated to incredible precision. Robot spacecraft carrying arrays of scientific instruments have been sent to rendezvous with them. Old, familiar theories of formation, evolution and impact have been revised and new ones presented. Facts abound. The data seems to change on an almost daily basis. I have attempted to compile much of what I consider "important," adding to that some of what I would call "interesting." I apologize in advance for the inevitable bits of data which will change during the lifetime of this book.

The solar system.
Doppelmayr,
Johann Gabriel,
Atlas Coelestis,
Nuremberg, 1742.
(Linda Hall Library)

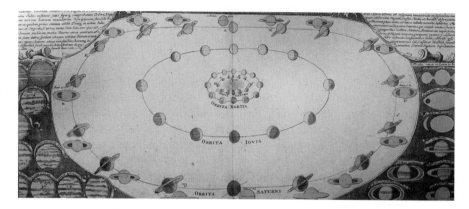

Thus, it is in a spirit of humility that I offer the present work. I now know that no book can hope to contain all the information available on the planets. Imagine this author in his study one year ago, thinking that after writing a book on the 88 constellations, *all I had to do* was another on *only* 9 planets. What was I thinking?

Acknowledgements

First and foremost, I wish to thank my lovely wife, Holley, for the love, the encouragement, the great attitude, and, of course, for originating all the non-photographic illustrations used in this book. Her talent and willingness to help – even in the wee hours of many mornings – was nothing short of a blessing.

Also, I wish to thank Ray and Carol Shubinski for a few ideas related to the narrative sections during a fun and productive drive to Las Cruces. Ray also gets credit for urging me – and it seems oh, so many years ago – to build up a credible astronomy reference library above and beyond my collection. That is one piece of advice I am glad to have followed.

Thanks to Bruce Bradley and Cindy Rogers at the Linda Hall Library in Kansas City, MO, for their help in securing many of the ancient photographs and drawings used herein. As the selected illustrations demonstrate, the Linda Hall Library continues to be one of the premier repositories of science literature on the planet. The collections at Linda Hall were also used extensively in the research for this book. Thanks also to the people at the Space Telescope Science Institute. Material created with support to AURA/ST ScI from NASA contrace NAS5-26555 is reproduced here with permission.

As with my first book, I would like to thank the team at Cambridge University Press. A heartfelt "thank you" to Simon Mitton who devoted more than a little time to this project to help me fashion it into a work that, I hope, is worthy of the Cambridge name. Thanks also to Alice Houston who helped in the formative early stages of the project. And the biggest "thank-you" to Jo Clegg, who, as with my last book, devoted so much time and careful effort into shaping the manuscript into what you see before you. Quite simply, Jo, you are the best!

Part 1

The Lists

Planets

Albedo

Mercury	0.10
Venus	0.65
Earth	0.37
Mars	0.15
Jupiter	0.52
Saturn	0.47
Uranus	0.51
Neptune	0.41
Pluto	0.30

Angular size

	Angular size[a]	
	Maximum (equatorial)	Minimum (equatorial)
Mercury	10″	4.9″
Venus	64″	10″
Mars	25.16″	3.5″
Jupiter [b]	50.11″	30.467″
Saturn [c]	20.75″	18.44″
Uranus [d]	3.96″	3.60″
Neptune	2.52″	2.49″
Pluto [e]	0.11″	0.065″

Notes:
[a] As seen from Earth.
[b] Jupiter's polar diameter is 93% of its equatorial diameter.
[c] Saturn's polar diameter is 89% of its equatorial diameter. The diameter of its rings is 225% of its equatorial diameter.
[d] Uranus's polar diameter is 97% of its equatorial diameter.
[e] Pluto reached perihelion 4 Sep 1989 and closest approach to Earth 7 May 1990. At that moment, its disk diameter was at a maximum. Its size is now shrinking and will continue to decrease each year – ever so slightly – until 2115, when Pluto's apparent size will again begin to grow.

Atmospheric composition

Note: $0.0001\% = 1$ ppm (part per million).

Mercury	Potassium (K)	31.7%
	Sodium (Na)	24.9%
	Oxygen (O)	9.5%
	Argon (Ar)	7.0%
	Helium (He)	5.9%
	Molecular oxygen (O_2)	5.6%
	Molecular nitrogen (N_2)	5.2%
	Carbon dioxide (CO_2)	3.6%
	Water vapor (H_2O)	3.4%
	Molecular hydrogen (H_2)	3.2%
Venus	Carbon dioxide (CO_2)	>96.4%
	Molecular nitrogen (N_2)	>3.4%
	Sulfur dioxide (SO_2)	0.015%
	Argon (Ar)	0.007%
	Water vapor (H_2O)	0.002%
	Carbon monoxide (CO)	0.0017%
	Helium (He)	0.0012%
	Neon (Ne)	0.0007%
	Carbonyl sulfide (COS)	0.00044%
	Hydrogen chloride (HCl)	0.00004%
	Hydrogen fluoride (HF)	0.000001%
Earth	Molecular nitrogen (N_2)	>78.0%
	Molecular oxygen (O_2)	>20.9%
	Argon (Ar)	0.93%
	Carbon dioxide (CO_2)	0.03%
	Neon (Ne)	0.002%
Mars	Carbon dioxide (CO_2)	>95.3%
	Molecular nitrogen (N_2)	>2.6%
	Argon (Ar)	1.6%
	Molecular oxygen (O_2)	0.13%
	Carbon monoxide (CO)	0.07%
	Water vapor (H_2O)	0.03%
	Neon (Ne)	0.00025%
	Krypton (Kr)	0.00003%
	Xenon (Xe)	0.000008%
	Ozone (O_3)	0.000003%

Atmospheric composition

Jupiter Molecular hydrogen (H_2) >81%
Helium (He) >17%
Methane (CH_4) 0.1%
Water vapor (H_2O) 0.1%
Ammonia (NH_3) 0.02%
Ethane (C_2H_6) 0.0002%
Phosphine (PH_3) 0.0001%
Hydrogen sulfide (H_2S) <0.0001%
Acetylene (C_2H_2) 0.000003%
Monodeuteromethane (CH_3D) 0.000002%
Hydrogen cyanide (HCN) 0.0000001%
Ethylene (C_2H_4) 0.0000001%
Methylamine (CH_3NH_2) <0.0000001%
Hydrazine (N_2H_4) <0.0000001%
Germane (GeH_4) 0.00000006%
Carbon monoxide (CO) 0.00000001%

Saturn Molecular hydrogen (H_2) >93%
Helium (He) >5%
Methane (CH_4) 0.2%
Water vapor (H_2O) 0.1%
Ammonia (NH_3) 0.01%
Ethane (C_2H_6) 0.0005%
Phosphine (PH_3) 0.0001%
Hydrogen sulfide (H_2S) <0.0001%
Methylamine (CH_3NH_2) <0.0001%
Acetylene (C_2H_2) 0.00001%
Monodeuteromethane (CH_3D) 0.000002%
Hydrogen cyanide (HCN) 0.0000001%
Hydrazine (N_2H_4) <0.0000001%
Germane (GeH_4) <0.0000001%
Ethylene (C_2H_4) <0.0000001%
Carbon monoxide (CO) <0.00000001%

Uranus Molecular hydrogen (H_2) >82%
Helium (He) >14%
Methane (CH_4) 2%
Ammonia (NH_3) 0.01%
Ethane (C_2H_6) 0.00025%
Acetylene (C_2H_2) 0.00001%
Carbon monoxide (CO) <0.00000001%
Hydrogen sulfide (H_2S) <0.00000001%

Atmospheric composition

Neptune	Molecular hydrogen (H_2)	$>84\%$
	Helium (He)	$>12\%$
	Methane (CH_4)	2%
	Ammonia (NH_3)	0.01%
	Ethane (C_2H_6)	0.00025%
	Acetylene (C_2H_2)	0.00001%
	Carbon monoxide (CO)	$<0.00000001\%$
	Hydrogen sulfide (H_2S)	$<0.00000001\%$
Pluto	Methane (CH_3)	Only compound detected

Atmospheric pressure

Mercury None

Venus 9 321 900 pascals

Earth Varies from 101 325 pascals at sea level

Mars 699 to 912 pascals

Jupiter Varies with depth to greater than 10 132 500 pascals

Saturn Varies with depth to greater than 10 132 500 pascals

Uranus Varies with depth to greater than 10 132 500 pascals

Neptune Varies with depth to greater than 10 132 500 pascals

Pluto None

Brightness and size of the Sun from each planet

	Size	Brightness[a]
Mercury	1.38°	−28.77
Venus	0.74°	−27.35
Earth	0.53°	−26.7
Mars	0.35°	−25.8
Jupiter	0.102°	−23.1
Saturn	0.056°	−21.8
Uranus	0.028°	−20.3
Neptune	0.017°	−19.1
Pluto	0.013°	−18.5

Note:
[a] Values given are the most accurate available at the time of writing.

Brilliancy at opposition

Brilliancy[a,b]

	Maximum visual magnitude	Minimum visual magnitude
Mars	−2.9	−1.0
Jupiter	−2.9	−2.0
Saturn	−0.3	+0.9
Uranus	+5.65	+6.06
Neptune	+7.66	+7.70
Pluto	+13.6	+15.95

Notes:
[a] Values given are the most accurate available at the time of writing.
[b] Mercury and Venus are never in opposition.

Cloud features

Mercury   None.

Venus   There are no permanent visible light features in the clouds of Venus. The atmosphere is in a continuous state of mixing, and any patterns observed quickly dissipate.

Earth   No permanent features are visible in the atmosphere of Earth. Occasional storms (such as hurricanes) are visible from space.

Mars   No permanent features are visible in the atmosphere of Mars, as the air is very thin. Occasionally, dust storms may be visible as they obscure surface features.

Jupiter   The Great Red Spot, a storm 30 000 km in length, dominates the visible atmosphere of Jupiter. The Red Spot has been observed for over 100 years, and may well be a permanent feature in Jupiter's upper cloud deck, which is composed primarily of ammonia crystals. In addition to the Great Red Spot, a wide variety of bands in the polar, temperate and equatorial regions are always visible.

Saturn   Although similar in composition and structure to that of Jupiter, the visible cloud deck of Saturn is much more restrained in appearance. Still, a number of cloud bands are usually visible along with temporary features such as spots.

Uranus   The atmosphere of Uranus is remarkably featureless due to the lack of an internal planetary heat source. Such a source, as is found in the atmosphere of Jupiter (and, to a lesser extent, Saturn), drives convective activity. The upper cloud deck of Uranus is composed of a methane haze, rather than the ammonia which dominates the two giant planets.

Neptune   The atmosphere of Neptune is similar to that of Uranus in composition and temperature. However, that is where the similarity ends. At a depth where the pressure is 101 325 pascals methane dominates as the principal component. The atmosphere above this is essentially transparent, unlike the hazy layer of Uranus. Somewhat deeper into the atmosphere, where the pressure is approximately 303 975 pascals, another cloud layer exists, composed primarily of hydrogen sulfide ice crystals.

Pluto   None.

Constellations visited by the Moon and planets

In addition to the 12 zodiacal constellations, the Moon and planets may appear in the following constellations.

The Moon

Auriga	Cetus	Corvus
Ophiuchus	Orion	Sextans

Total = 18 (counting the 12 traditional zodiacal constellations)

The visible planets

(in addition to the constellations listed above)

Canis Minor	Crater	Hydra
Pegasus	Scutum	Serpens

Total = 24

Pluto

(in addition to the constellations listed above)

Andromeda	Aquila	Boötes
Centaurus	Coma Berenices	Corona Australis
Equuleus	Eridanus	Leo Minor
Lupus	Lynx	Microscopium
Monoceros	Perseus	Piscis Austrinus
Sculptor	Triangulum	

Total = 41

Density

	Density (g/cm^3)[a]	Ratio (Earth = 1)
Mercury	5.43	0.982 77
Venus	5.25	0.951 95
Earth	5.515	1.000 00
Mars	3.94	0.714 42
Jupiter	1.33	0.241 16
Saturn	0.69	0.125 11
Uranus	1.29	0.233 91
Neptune	1.64	0.297 37
Pluto	2.05	0.371 71

Note:
[a] Values given are the most accurate available at the time of writing.

Distances from Earth

Distance from Earth

	Maximum	Minimum
Mercury	221 920 880 km 1.48 AU	77 269 900 km 0.52 AU
Venus	261 039 880 km 1.75 AU	38 150 900 km 0.26 AU
Mars	401 355 980 km 2.68 AU	54 510 620 km 0.36 AU
Jupiter	968 460 580 km 6.47 AU	588 404 520 km 3.93 AU
Saturn	1 658 854 980 km 11.09 AU	1 195 772 020 km 7.99 AU
Uranus	3 159 769 980 km 21.12 AU	2 582 694 020 km 17.26 AU
Neptune	4 686 510 980 km 31.33 AU	4 306 660 020 km 28.79 AU
Pluto	7 676 691 980 km 51.30 AU	4 283 023 020 km 28.63 AU

Distances from Sun

Distance from Sun

	Maximum	Average	Minimum
Mercury	69 815 900 km 0.4667 AU	57 910 000 km 0.3871 AU	46 003 500 km 0.3075 AU
Venus	108 934 900 km 0.7282 AU	108 200 000 km 0.7233 AU	107 463 300 km 0.7184 AU
Earth	152 104 980 km 1.0168 AU	149 597 870 km 1.0000 AU	147 085 800 km 0.9832 AU
Mars	249 251 000 km 1.6661 AU	227 940 000 km 1.5237 AU	206 615 600 km 1.3811 AU
Jupiter	816 355 600 km 5.4570 AU	778 330 000 km 5.2028 AU	740 509 500 km 4.9500 AU
Saturn	1 506 750 000 km 10.0720 AU	1 429 400 000 km 9.5388 AU	1 347 877 000 km 9.0100 AU
Uranus	3 007 665 000 km 20.1050 AU	2 870 990 000 km 19.1914 AU	2 734 799 000 km 18.2810 AU
Neptune	4 534 406 000 km 30.3106 AU	4 504 300 000 km 30.0611 AU	4 458 765 000 km 29.8050 AU
Pluto	7 524 587 000 km 50.2987 AU	5 913 520 000 km 39.5294 AU	4 435 128 000 km 29.6470 AU

Eccentricity

Mercury	0.2056
Venus	0.0068
Earth	0.0167
Mars	0.0934
Jupiter	0.0483
Saturn	0.0560
Uranus	0.0461
Neptune	0.0097
Pluto	0.2482

Escape velocity

	Escape velocity[a]		Ratio (Earth = 1)
	km/s	km/hr	
Mercury	4.25	(15 300)	0.380
Venus	10.36	(37 296)	0.927
Earth	11.18	(40 248)	1.000
Mars	5.02	(18 072)	0.449
Jupiter	59.366	(213 718)	5.310
Saturn	35.49	(127 764)	3.174
Uranus	21.30	(76 680)	1.905
Neptune	23.50	(84 600)	2.102
Pluto	1.22	(4392)	0.109

Note:
[a] Values given are the most accurate available at the time of writing.

Future dates of conjunction

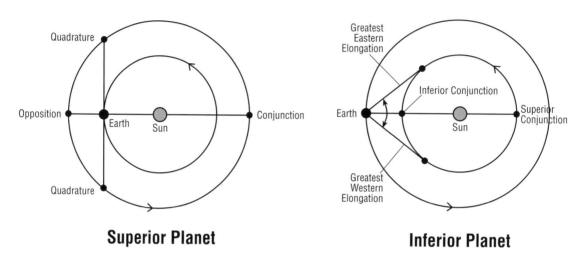

Superior Planet

Inferior Planet

Planetary configurations. (Holley Bakich)

Inferior planets   Dates of conjunction

Inferior	Superior
Mercury	

Inferior	Superior
26 Jul 1999	8 Sep 1999
15 Nov 1999[a]	16 Jan 2000
1 Mar 2000	9 May 2000[a]
6 Jul 2000	22 Aug 2000
30 Oct 2000	25 Dec 2000
13 Feb 2001	23 Apr 2001
16 Jun 2001	5 Aug 2001
14 Oct 2001	4 Dec 2001
27 Jan 2002	7 Apr 2002
27 May 2002	21 Jun 20021
27 Sep 2002	4 Nov 2002[a]
11 Jan 2003	21 Mar 2003
7 May 2003[a]	5 Jul 2003
11 Sep 2003	25 Oct 2003
27 Dec 2003	4 Mar 2004
17 Apr 2004	18 Jun 2004
23 Aug 2004	5 Oct 2004
10 Dec 2004	14 Feb 2005

Inferior	Superior
29 Mar 2005	3 Jun 2005
5 Aug 2005	18 Sep 2005
24 Nov 2005	27 Jan 2006
12 Mar 2006	19 May 2006
18 July 2006	1 Sep 2006
9 Nov 2006	7 Jan 2007
23 Feb 2007	3 May 2007
28 Jun 2007	16 Aug 2007
24 Oct 2007	17 Dec 2007
6 Feb 2008	16 Apr 2008
7 Jun 2008	29 Jul 2008
7 Oct 2008	25 Nov 2008
20 Jan 2009	31 Mar 2009
18 May 2009	14 Jul 2009
20 Sep 2009	5 Nov 2009
4 Jan 2010	14 Mar 2010
29 Apr 2010	28 Jun 2010
3 Sep 2010	17 Oct 2010
20 Dec 2010	

	Inferior	Superior
Venus	20 Aug 1999	11 Jun 2000
	30 Mar 2001	14 Jan 2002
	31 Oct 2002	18 Aug 2003
	8 Jun 2004[a]	31 Mar 2005
	14 Jan 2006	28 Oct 2006
	18 Aug 2007	9 Jan 2008
	28 Mar 2009	11 Jan 2010
	29 Oct 2010	

Note:
[a] Transit.

21

Superior planets Dates of conjunction

Mars 1 Jul 2000
10 Aug 2002
15 Sep 2004
24 Oct 2006
6 Dec 2008
4 Feb 2011

Jupiter 8 May 2000
14 Jun 2001
20 Jul 2002
22 Aug 2003
21 Sep 2004
22 Oct 2005
22 Nov 2006
23 Dec 2007
24 Jan 2009
28 Feb 2010

Saturn 10 May 2000
25 May 2001
9 Jun 2002
24 Jun 2003
8 Jul 2004
23 Jul 2005
7 Aug 2006
22 Aug 2007
4 Sep 2008
17 Sep 2009
1 Oct 2010

Uranus 6 Feb 2000
9 Feb 2001
13 Feb 2002
17 Feb 2003
22 Feb 2004
25 Feb 2005
1 Mar 2006
5 Mar 2007
9 Mar 2008
13 Mar 2009
17 Mar 2010

Neptune	24 Jan 2000
	26 Jan 2001
	28 Jan 2002
	30 Jan 2003
	2 Feb 2004
	3 Feb 2005
	6 Feb 2006
	8 Feb 2007
	11 Feb 2008
	12 Feb 2009
	15 Feb 2010

Pluto	3 Dec 1999
	4 Dec 2000
	7 Dec 2001
	9 Dec 2002
	12 Dec 2003
	13 Dec 2004
	16 Dec 2005
	18 Dec 2006
	21 Dec 2007
	22 Dec 2008
	24 Dec 2009
	27 Dec 2010

Future dates of opposition

Dates of opposition[a]

	Date	Constellation
Mars	13 Jun 2001	Ophiuchus
	28 Aug 2003	Aquarius
	7 Nov 2005	Aries
	24 Dec 2007	Gemini
	29 Jan 2010	Cancer
Jupiter	23 Oct 1999	Aries
	28 Nov 2000	Taurus
	1 Jan 2002	Gemini
	2 Feb 2003	Cancer
	4 Mar 2004	Leo
	3 Apr 2005	Virgo
	4 May 2006	Libra
	6 Jun 2007	Ophiuchus
	9 Jul 2008	Sagittarius
	15 Aug 2009	Capricornus
	21 Sep 2010	Pisces
Saturn	6 Nov 1999	Aries
	19 Nov 2000	Taurus
	3 Dec 2001	Taurus
	17 Dec 2002	Taurus
	31 Dec 2003	Gemini
	13 Jan 2005	Gemini
	28 Jan 2006	Cancer
	10 Feb 2007	Leo
	24 Feb 2008	Leo
	8 Mar 2009	Leo
	22 Mar 2010	Virgo

	Date	Constellation
Uranus	7 Aug 1999	Capricornus
	11 Aug 2000	Capricornus
	15 Aug 2001	Capricornus
	20 Aug 2002	Capricornus
	24 Aug 2003	Aquarius
	27 Aug 2004	Aquarius
	1 Sep 2005	Aquarius
	5 Sep 2006	Aquarius
	9 Sep 2007	Aquarius
	13 Sep 2008	Aquarius
	17 Sep 2009	Pisces
	21 Sep 2010	Pisces
Neptune	26 Jul 1999	Capricornus
	27 Jul 2000	Capricornus
	30 Jul 2001	Capricornus
	2 Aug 2002	Capricornus
	4 Aug 2003	Capricornus
	6 Aug 2004	Capricornus
	8 Aug 2005	Capricornus
	11 Aug 2006	Capricornus
	13 Aug 2007	Capricornus
	15 Aug 2008	Capricornus
	18 Aug 2009	Capricornus
	20 Aug 2010	Capricornus
Pluto	1 Jun 2000	Ophiuchus
	4 Jun 2001	Ophiuchus
	7 Jun 2002	Ophiuchus
	9 Jun 2003	Ophiuchus
	11 Jun 2004	Ophiuchus
	14 Jun 2005	Serpens
	17 Jun 2006	Serpens
	19 Jun 2007	Sagittarius
	21 Jun 2008	Sagittarius
	23 Jun 2009	Sagittarius
	25 Jun 2010	Sagittarius

Note:
[a] Mercury and Venus are never in opposition.

Future significant alignments

Significant alignments[a]

Date	Separation in declination
Mercury with Venus	
28 Apr 2000	21'
21 Jun 2003	25'
14 Jan 2005	21'
27 Jun 2005	05'
with Mars	
19 May 2000	67'
10 Aug 2000	05'
25 Jul 2002	40'
10 Jul 2004	10'
29 Sep 2004	51'
with Jupiter	
8 May 2000	52'
20 Jul 2002	75'
26 Jul 2003	23'
28 Sep 2004	40'
6 Oct 2005	88'
with Saturn	
2 Jul 2002	14'
1 Jul 2003	93'
26 Jun 2006	85'

	Date	Separation in declination
Venus	with Mercury	
	28 Apr 2000	21′
	21 Jun 2003	25′
	14 Jan 2005	21′
	27 Jun 2005	05′
	with Mars	
	21 Jun 2000	18′
	10 May 2002	18′
	5 Dec 2004	75′
	with Jupiter	
	17 May 2000	01′
	5 Aug 2001	72′
	3 Jun 2002	99′
	21 Aug 2003	34′
	4 Nov 2004	36′
	2 Sep 2005	82′
	with Saturn	
	18 May 2000	74′
	15 Jul 2001	44′
	8 Jul 2003	49′
	25 Jun 2005	78′
Mars	with Mercury	
	19 May 2000	67′
	10 Aug 2000	05′
	25 Jul 2002	40′
	10 Jul 2004	10′
	29 Sep 2004	51′
	with Venus	
	21 Jun 2000	18′
	10 May 2002	18′
	5 Dec 2004	75′
	with Jupiter	
	6 Apr 2000	66′
	3 Jul 2002	49′
	27 Sep 2004	12′
	with Saturn	
	24 May 2004	95′

27

Significant alignments[a]

	Date	Separation in declination
Jupiter	with Mercury	
	8 May 2000	52'
	20 Jul 2002	75'
	26 Jul 2003	23'
	28 Sep 2004	40'
	6 Oct 2005	88'
	with Venus	
	17 May 2000	01'
	5 Aug 2001	72'
	3 Jun 2002	99'
	21 Aug 2003	34'
	4 Nov 2004	36'
	2 Sep 2005	82'
	with Mars	
	6 Apr 2000	66'
	3 Jul 2002	49'
	27 Sep 2004	12'
	with Saturn	
	31 May 2000	71'
Saturn	with Mercury	
	2 Jul 2002	14'
	1 Jul 2003	93'
	26 Jun 2006	85'
	with Venus	
	18 May 2000	74'
	15 Jul 2001	44'
	8 Jul 2003	49'
	25 Jun 2005	78'
	with Mars	
	24 May 2004	95'
	with Jupiter	
	31 May 2000	71'

Note:
[a] No alignments are provided for Uranus, Neptune or Pluto as Uranus normally requires a telescope to view it and Neptune and Pluto always require optical aid.

Future transits

Transits[a]

Date	Constellation	Ephemeris Time[b] of mid-transit
Mercury		
15 Nov 1999	Libra	21^h43^m
7 May 2003	Aries	07^h53^m
8 Nov 2006	Libra	21^h42^m
9 May 2016	Aries	15^h00^m
11 Nov 2019	Libra	15^h22^m
13 Nov 2032	Libra	08^h58^m
Venus		
8 Jun 2004	Taurus	08^h24^m
6 Jun 2012	Taurus	01^h36^m
11 Dec 2117	Ophiuchus	02^h51^m
8 Dec 2125	Ophiuchus	16^h01^m
11 Jun 2247	Taurus	11^h43^m
9 Jun 2255	Taurus	04^h50^m

Notes:
[a] Transits of the outer planets are not possible.
[b] All times are geocentric.

Inclination of orbit

Mercury	7.004°
Venus	3.394°
Earth	0.000°
Mars	1.850°
Jupiter	1.308°
Saturn	2.488°
Uranus	0.774°
Neptune	1.774°
Pluto	17.148°

Magnetic field strength and orientation

Mercury Mercury has slightly more than 1% of the magnetic field of the Earth. The fact that it has any magnetic field at all is amazing, considering the slowness of Mercury's rotation. The field is probably due to its composition. Mercury has an iron core, much like the Earth. The core of Mercury comprises approximately 70–80% of the planet's radius, the outer region being largely composed of silicate rocks.

Venus Venus has little or no magnetic field of its own. This is almost certainly due to the slowness of its rotation, rather than because of the internal composition of the planet itself. Venus probably does possess a liquid metal core, similar to the Earth's. Separate from planetary magnetism, a sizable magnetic field is induced in the upper atmosphere (ionosphere) by the solar wind.

Earth The magnetic field of the Earth results from electric currents circulating within the liquid core. Planetary rotation induces a dynamo-like effect in the metallic core, caused by the slow motion of the materials there. The strength and alignment of the magnetic field vary due to changes in these motions. The extent of the magnetic field into space also changes, but generally it extends 48 000–64 000 km above the Earth's surface in the direction toward the Sun and 322 000–362 000 km in the direction opposite the Sun. The difference is due to the pressure exerted on the magnetosphere by the solar wind.

Details of Earth's magnetic field

Average surface field (tesla)	0.000 03
Dipole moment (weber-meters)	6×10^{14}
Tilt from planetary axis	11°
Offset from planetary axis (planet radii)	0.0

Mars Mars has little or no magnetic field due to the fact that, unlike the Earth, the core of the planet is not liquid metal. Mars, like the Moon, lacks sufficient mass to produce the pressure (and, therefore, the temperature) associated with maintaining a liquid core. This reasoning makes the magnetic field of Mercury all the more remarkable.

Jupiter Reaching a distance of more than 1 600 000 km from the planet, the magnetosphere of Jupiter is the most extensive of any of the planets. The strong field is generated by Jupiter's rapid rotation coupled with an interior of metallic liquid hydrogen, which acts in a way similar to the liquid iron core of the Earth.

Details of Jupiter's magnetic field

Average surface field (tesla)	0.0004
Dipole moment (weber-meters)	3.14×10^{20}
Tilt from planetary axis	10°
Offset from planetary axis (planet radii)	0.0

Saturn As is the case with Jupiter, the strong field is generated by Saturn's rotation – nearly as fast as Jupiter's – along with a similar interior of metallic liquid hydrogen.

Details of Saturn's magnetic field

Average surface field (tesla)	0.00002
Dipole moment (weber-meters)	3×10^{17}
Tilt from planetary axis	1°
Offset from planetary axis (planet radii)	0.0

Uranus Rapid rotation certainly aids the generation of a sizable magnetic field for Uranus. Unlike Jupiter and Saturn, both of whose metallic regions are near their respective cores, the metallic region of Uranus seems to be in the mantle of the planet. It has been suggested that this is the reason for the large offset of the magnetic field from the planet's core. The size of the magnetosphere of Uranus varies from 15–20 times the radius of the planet. Its composition is primarily hydrogen ions originating from the solar wind and from the atmosphere of Uranus. The bow shock of Uranus was encountered by Voyager 2 at a distance of 23.7 Uranian radii from the planet and the magnetopause was crossed (on the sunward side) at a distance of 18 Uranian radii.

Details of Uranus's magnetic field

Average surface field (tesla)	0.00003
Dipole moment (weber-meters)	3×10^{16}
Tilt from planetary axis	59°
Offset from planetary axis (planet radii)	0.3

Neptune   Neptune's magnetic field is almost an exact duplicate of that of Uranus in size, strength, inclination to planetary axis and offset from the center of the planet. The magnetosphere, like that of Uranus, is composed of hydrogen ions that originate from Neptune's atmosphere and from the solar wind. Neptune's bow shock was encountered by Voyager 2 at 34.9 Neptunian radii, and a "fuzzy" magnetosphere was crossed near 26 Neptunian radii.

Details of Neptune's magnetic field

Average surface field (tesla)	0.00002
Dipole moment (weber-meters)	2×10^{16}
Tilt from planetary axis	55°
Offset from planetary axis (planet radii)	0.5

Pluto   At the time of writing, no magnetic field associated with Pluto has been detected. Pluto almost certainly lacks the mass to maintain a liquid metal core, and the slowness of its rotation would diminish the possibility of a magnetic field.

Mass

	Mass (kg)	Ratio (Earth = 1)
Mercury	3.303×10^{23}	0.055
Venus	4.869×10^{24}	0.815
Earth	5.976×10^{24}	1.000
Mars	6.421×10^{23}	0.107
Jupiter	1.900×10^{27}	317.940
Saturn	5.688×10^{26}	95.181
Uranus	8.686×10^{25}	14.535
Neptune	1.024×10^{26}	17.135
Pluto	1.290×10^{22}	0.002

Named features on the planets and the Moon

The International Astronomical Union (IAU) has been the arbiter of planetary and satellite nomenclature since its organizational meeting in 1919. At that time a committee was appointed to standardize the varied lunar and Martian nomenclatures then in use.

The report of this committee, *Named Lunar Formations* by M. Blagg and K. Müller (1935), was the first systematic listing of lunar nomenclature. Later, *The System of Lunar Craters, quadrants I, II, III, IV* was published by D. W. G. Arthur and others (1963, 1964, 1965, 1966), under the direction of Gerard P. Kuiper. The accompanying maps (also in four parts) became the recognized sources for lunar nomenclature.

The Martian nomenclature was clarified in 1958, when a committee of the IAU recommended adopting the names of 128 features observed through ground-based telescopes. These names were based on a system of nomenclature developed by the Italian astronomer G. V. Schiaparelli (1879) and expanded by E. M. Antoniadi (1929), a Turkish-born astronomer working at Meudon, France.

In 1970, a Mars nomenclature working group was formed. At about the same time, names of features on the Moon were updated as a lunar committee was formed that suggested names for features discriminated by the Soviet Zond and American Lunar Orbiter and Apollo cameras.

In 1973, a new group, the Working Group for Planetary System Nomenclature was appointed. Task groups for the Moon and planets were formed to conduct the preliminary work of choosing themes and proposing names for newly discriminated features. A task group was formed in 1984 to name surface features on asteroids and comets.

When images are first obtained of the surface of a planet or satellite, a theme for naming features is chosen and a few important features are named, usually by members of the appropriate IAU task group. Later, as higher resolution images and maps become available, additional features are named, usually at the request of investigators mapping or describing specific surfaces, features or geologic formations.

Mercury Total named features: 297

Feature	Number	Theme for names
Craters	239	Famous deceased artists, musicians, painters, authors
Montes	1	Caloris, from Latin word for "hot"
Planitiae	7	Names for Mercury in various languages
Rupes	16	Ships of discovery or scientific expeditions
Valles	4	Radio telescope facilities
Albedo features	30	

Venus Total named features: 1761

Feature	Number	Theme for names
Chasmata	56	Goddesses of hunt; moon goddesses
Colles	12	Sea goddesses
Coronae	240	Fertility and earth goddesses
Craters	870	Over 20 km, famous women; under 20 km, common female first names
Dorsa	94	Sky goddesses
Farrum	9	Water goddesses
Fluctus	35	Goddesses, miscellaneous
Fossae	31	Goddesses of war
Lineae	16	Goddesses of war
Montes	100	Goddesses, miscellaneous (also one radar scientist)
Paterae	70	Famous women
Planitiae	41	Mythological heroines
Planum	2	Lakshmi; goddess of prosperity
Regiones	21	Giantesses and Titanesses (also Greek alphanumeric)
Rupes	7	Goddesses of hearth and home
Terrae	3	Goddesses of love
Tesserae	62	Goddesses of fate and fortune
Tholi	30	Goddesses, miscellaneous
Undae	3	Desert goddesses
Valles	59	Word for planet Venus in various world languages

Moon Total named features: 1940

Feature	Number	Theme for names
Albedo feature	1	Named from nearby crater
Apollo landing sites	78	Named by astronauts
Catenae	20	Named from nearby craters
Craters	1545	Large craters: famous deceased scientists, scholars, artists; small craters: common first names
Dorsa	40	Named from nearby craters
Lacus	20	Latin terms for weather[a]
Maria	22	Latin terms for weather[a]
Montes	51	Terrestrial mountain ranges or nearby craters
Oceanus	1	"Ocean of Storms"
Paludes	3	Latin terms for weather[a]
Planitia	1	Luna 9 landing site
Promontoria	9	Terrestrial capes and famous deceased scientists
Rimae	118	Named from nearby craters
Rupes	8	Named from nearby mountain ranges
Sinus	11	Latin terms for weather[a]
Valles	12	Named from nearby features

Note:
[a] And other abstract concepts.

Mars   Total named features: 1345

Feature	Number	Theme for names
Albedo features	127	From nearest named albedo feature on Schiaparelli or Antoniadi maps
Catenae	17	
Cavus	12	
Chaos	19	
Chasmata	23	
Colles	13	
Craters	845	Large craters: deceased scientists who have contributed to the study of Mars; small craters: villages of the world (less than 100000 population, UN Yearbook)
Dorsa	30	From nearest named albedo feature on Schiaparelli or Antoniadi maps
Fossae	56	
Labes	5	
Labyrinthus	3	
Mensae	21	
Montes	34	
Paterae	17	
Planitiae	11	
Planam	21	
Rupes	22	
Scopulus	11	
Sulcis	12	
Terrae	12	
Tholis	17	
Undae	2	
Valles	14	Large valles: name for Mars/star in various languages; small valles: classical or modern names of rivers
Vastitas	1	From nearest named albedo feature

Phobos:

Craters	2	Scientists who helped discovery

Deimos:

Craters	7	Authors who wrote about satellites
Dorsa	1	

Jupiter Total named features on satellites: 530

Feature	Number	Theme for names
Amalthea		
Craters	2	None
Faculae	2	
Io		
Active eruptive centers	11	Fire, sun and thunder gods and heroes
Catenae	2	Sun gods
Fluctus	9	Name derived from nearby named feature, or fire, sun, thunder and volcano gods, goddesses and heroes, mythical blacksmiths
Mensae	5	People associated with Io myth, derived from nearby feature, or from Dante's *Inferno*
Montes	14	Places associated with Io myth, derived from nearby feature, or from Dante's *Inferno*
Paterae	100	Fire, sun, thunder and volcano gods, heroes and goddesses, mythical blacksmiths
Plana	8	Places associated with Io myth, derived from nearby feature, or from Dante's *Inferno*
Regiones	9	
Tholi	2	
Europa:		
Chaos	1	Places associated with Celtic myths
Craters	8	Celtic gods and heroes
Flexus	5	Places associated with the Europa myth
Large ring features	2	Celtic stone circles
Lenticulae	—	Celtic gods and heroes
Lineae	30	People associated with the Europa myth
Maculae	5	Places associated with the Europa myth
Regiones	—	Places associated with Celtic myths

Jupiter cont.	Feature	Number		Theme for names
	Ganymede:			
	Catenae	3	}	Gods and heroes of ancient Fertile Crescent people
	Craters	111		
	Faculae	14		Places associated with Egyptian myths
	Fossae	3		Gods (or principals) of ancient Fertile Crescent people
	Regiones	5		Astronomers who discovered Jovian satellites
	Sulci	29		Places associated with myths of ancient people
	Callisto:			
	Catenae	8		Mythological places in high latitudes
	Craters	139		Heroes and heroines from northern myths
	Large ring features	3		Homes of the gods and of heroes

Saturn Total named features on satellites: 191

Feature	Number	Theme for names
Hyperion:		
Craters	4	} Sun and Moon deities
Dorsi	1	
Epimetheus:		
Craters	2	People from myth of Castor and Pollux
Mimas:		
Chasmata	7	} People and places from Malory's *Le Morte*
Craters	29	*d'Arthur* legends (Baines translation)[a]
Enceladus:		
Craters	15	
Fossae	3	People and places from Burton's *Arabian*
Planitiae	2	*Nights*
Sulci	2	
Tethys:		
Chasmata	1	} People and places from Homer's *Odyssey*
Craters	19	
Dione:		
Chasmata	4	
Craters	25	People and places from Virgil's *Aeneid*
Lineae	3	
Rhea:		
Chasmata	2	} People and places from creation myths
Craters	48	
Janus:		
Craters	4	People from myth of Castor and Pollux
Iapetus:		
Craters	18	
Regiones	1	People and places from Sayers' translation
Terrae	1	of *Chanson de Roland*

Note:
[a] Except for the crater named Herschel.

Uranus Total named features on satellites: 88

Feature	Number	Theme for names
Puck		
Craters	3	Mischievous (Pucklike) spirits (class)
Miranda		
Craters	7	
Coronae	3	
Regiones	4	Characters and places from
Rupes	2	Shakespeare's plays
Sulci	2	
Ariel		
Chasmata	7	
Craters	17	Light spirits (individual and class)
Valles	2	
Umbriel		
Craters	13	Dark spirits (individual)
Titania		
Chasmata	2	
Craters	15	Shakespearean places and female
Rupes	1	characters
Oberon		
Chasmata	1	Shakespearean tragic heroes and places
Small satellites		
All features	0	Heroines from Shakespeare and Pope

Neptune   Total named features on satellites: 62

Feature	Number	Theme for names
Proteus		
Craters	1	Water-related spirits, gods, goddesses (excluding Greek and Roman names)
Nereid		
All features	0	Individual nereids
Triton		
Catenae	2	
Cavus	10	
Craters	9	
Dorsum	1	Aquatic names, excluding Roman and
Fossae	3	Greek. Possible categories include
Maculae	7	worldwide aquatic spirits, famous
Paterae	5	terrestrial fountains or fountain
Plana	3	locations, terrestrial aquatic features,
Planitiae	4	famous terrestrial geysers or geyser
Plumes	2	locations, terrestrial islands
Regiones	3	
Sulci	12	
Small satellites		
All features	0	Gods and goddesses associated with Neptune/Poseidon mythology or generic mythological aquatic beings

Pluto   There are no named features on Pluto or its satellite.

Names of the planets, Sun and Moon around the world

	Arabic	Danish	French	German	Greek	Hebrew
Sun	Shams	Solen	Soleil	Sonne	Helios	Shemesh
Mercury	Otaared	Merkur	Mercure	Merkur	Hermes	Kokhav Khama
Venus	Zuhra	Venus	Vénus	Venus	Aphrodite	Nogah
Earth	Ard	Jorden	Terre	Erde	Gea	Arets
Moon	Quamar	Månen	Lune	Mond	Selene	Yareakh
Mars	Merrikh	Mars	Mars	Mars	Ares	Ma'adim
Jupiter	Mushtarie	Jupiter	Jupiter	Jupiter	Zeus	Tzedek
Saturn	Zuhal	Saturn	Saturne	Saturn	Kronos	Shabtay
Uranus	Uraanus	Uranus	Uranus	Uranus	Uranos	Uranus
Neptune	Niptuun	Neptun	Neptune	Neptun	Poseidon	Neptune
Pluto	Plutoon	Pluto	Pluton	Pluto	Pluton	Pluto

	Italian	Japanese	Latin	Russian	Spanish	Swedish
Sun	Sole	Taiyou	Sol	Solnce	Sol	Solen
Mercury	Mercurio	Suisei	Mercurius	Merkurij	Mercurio	Merkurius
Venus	Venere	Kinsei	Venus	Venera	Venus	Venus
Earth	Terra	Chikyu	Terra	Zemlja	Tierra	Jorden
Moon	Luna	Tsuki	Luna	Luna	Luna	Månen
Mars	Marte	Kasei	Mars	Mars	Marte	Mars
Jupiter	Giove	Mokusei	Jupiter	Yupiter	Júpiter	Jupiter
Saturn	Saturno	Dosei	Saturnus	Saturn	Saturno	Saturnus
Uranus	Urano	Tennousei	Uranus	Uran	Urano	Uranus
Neptune	Nettuno	Kaiousei	Neptunus	Neptun	Neptuno	Neptunus
Pluto	Plutone	Meiousei	Pluto	Pluton	Pluton	Pluto

Oblateness

Note: The term "oblateness" is sometimes referred to as a planet's "ellipticity."

	Oblateness[a]
Mercury[b]	0
Venus[b,c]	0
Earth[d]	0.003 35
Mars	0.005 19
Jupiter	0.064 81
Saturn	0.1076
Uranus	0.030
Neptune	0.026
Pluto[b]	0

Notes:
[a] Values given are the most accurate available at the time of writing.
[b] There is no appreciable difference between the equatorial and polar diameters of Mercury, Venus and Pluto. This is almost certainly due to the slowness of rotation of both planets.
[c] The ratio of Venus's equatorial diameter to that of Earth is 0.949. The ratio of Venus's polar diameter to that of Earth is 0.952.
[d] Earth's polar diameter is 44 km less than its equatorial diameter.

Orbital period

Orbital period[a]

	Years	Days	Years–days–hours
Mercury	0.24085	87.969	$0^y87^d23.3^h$
Venus	0.61521	224.701	$0^y224^d16.8^h$
Earth	1.0000	365.2422	$1^y0^d0^h$
Mars	1.8809	686.98	$1^y320^d18.2^h$
Jupiter	11.8626	4332.71	$11^y315^d1.1^h$
Saturn	29.458	10759.3	$29^y167^d6.7^h$
Uranus	84.01	30684	$84^y3^d15.66^h$
Neptune	164.79	60188.3	$164^y288^d13^h$
Pluto	248.54	90777.3	$248^y197^d5.5^h$

Note:
[a] Values given are the most accurate available at the time of writing.

Orbital velocity

Orbital velocity

	m/s	km/s	km/hr
Mercury	47880	47.88	172368
Venus	35020	35.02	126072
Earth	29790	29.79	107244
Mars	24130	24.13	86868
Jupiter	13070	13.07	47052
Saturn	9670	9.67	34812
Uranus	6810	6.81	24516
Neptune	5450	5.45	19620
Pluto	4740	4.74	17064

Rotational period

Rotational period

	Days	Hours	Days–hours–minutes
Mercury	58.6462	1407.5088	$58^{d}15^{h}30.5^{m}$
Venus	243.0187 (retrograde)	5832.6088	$243^{d}0^{h}36.5^{m}$
Earth	0.99727	23.9345	$0^{d}23^{h}56.1^{m}$
Mars	1.025957	24.6240	$1^{d}0^{h}37.4^{m}$
Jupiter	0.41354	9.9250	$0^{d}9^{h}55.5^{m}$
Saturn	0.42637	10.2336	$0^{d}10^{h}14^{m}$
Uranus	0.71806	17.2344	$0^{d}17^{h}14^{m}$
Neptune	0.67125	16.1088	$0^{d}16^{h}6.5^{m}$
Pluto	6.3872	153.2928	$6^{d}9^{h}17.6^{m}$

Rotational velocity (equatorial)

	Rotational velocity (km/hr)[a]
Mercury	10.891
Venus	6.520
Earth	1 669.8
Mars	866.9
Jupiter	45 259.5
Saturn	37 004.9
Uranus	8971.5
Neptune	9667.1
Pluto	47.6

Note:
[a] Values given are the most accurate available at the time of writing.

Size

	Equatorial diameter (km)[a]	Polar diameter (km)[a]	Ratio (Earth = 1)[b]
Mercury	4879.4	4879.4	0.383
Venus	12104	12104	0.949
Earth	12756.28	12712	1.000
Mars	6794.4	6759	0.533
Jupiter	142984	133717	11.209
Saturn	120536	107566	9.449
Uranus	51118	49584	4.007
Neptune	49572	48283	3.886
Pluto	2320	2320	0.182

Notes:
[a] Values given are the most accurate available at the time of writing.
[b] The ratio is taken for the equatorial diameters.

Solar irradiance

	Solar irradiance (W/m²)[a]
Mercury	3566
Venus	2660
Earth	1380
Mars	595
Jupiter	51
Saturn	15
Uranus	3.8
Neptune	1.5
Pluto	0.9

Note:
[a] Values given are the most accurate available at the time of writing.

Speed of light travel times

	To or from the Sun			To or from the Earth	
	Maximum	Average	Minimum	Maximum	Minimum
Mercury	03^m53^s	03^m13^s	02^m33^s	12^m20^s	04^m18^s
Venus	06^m03^s	06^m01^s	05^m58^s	14^m31^s	02^m07^s
Earth	08^m27^s	08^m19^s	08^m11^s	—	—
Mars	13^m51^s	12^m40^s	11^m29^s	22^m19^s	03^m02^s
Jupiter	45^m23^s	43^m16^s	41^m10^s	53^m50^s	32^m43^s
Saturn	$01^h23^m46^s$	$01^h19^m28^s$	$01^h14^m56^s$	$01^h32^m13^s$	$01^h06^m28^s$
Uranus	$02^h47^m12^s$	$02^h39^m37^s$	$02^h32^m02^s$	$02^h55^m40^s$	$02^h23^m35^s$
Neptune	$04^h12^m05^s$	$04^h10^m25^s$	$04^h07^m53^s$	$04^h20^m33^s$	$03^h59^m25^s$
Pluto	$06^h58^m19^s$	$05^h28^m45^s$	$04^h06^m34^s$	$07^h06^m47^s$	$03^h58^m07^s$

Surface gravity

	Surface gravity (m/s^2)[a]	Ratio (Earth $= 1$)
Mercury	2.78	0.284
Venus	8.87	0.907
Earth	9.78	1.000
Mars	3.72	0.380
Jupiter	24.51[b]	2.506
Saturn	9.05[c]	0.925
Uranus	7.77	0.794
Neptune	11.0	1.125
Pluto	0.4	0.041

Notes:
[a] Values given are the most accurate available at the time of writing.
[b] This value is the equatorial surface gravity; the polar surface gravity is 26.36 m/s^2.
[c] This value is the equatorial surface gravity; the polar surface gravity is 10.14 m/s^2.

Synodic period

	Synodic period (days)
Mercury	115.88
Venus	583.92
Mars	779.94
Jupiter	398.88
Saturn	378.09
Uranus	369.66
Neptune	367.49
Pluto	366.73

Temperature range

	Temperate range (°C)		
	Minimum	Average	Maximum
Mercury	−173	179	427
Venus[a]	−44	464	500
Earth	−69	7	58
Mars	−140	−63	20
Jupiter	−163	−121	increases with depth
Saturn	−191	−130	increases with depth
Uranus	−214	−205	increases with depth
Neptune	−223	−220	increases with depth
Pluto	−240	−229	−218

Note:
[a] Minimum temperature occurs at the tops of the clouds of Venus.

Tilt of axis

Mercury	0.00°
Venus	177.36°
Earth	23.45°
Mars	25.19°
Jupiter	3.13°
Saturn	25.33°
Uranus	97.86°
Neptune	28.31°
Pluto	122.52°

Volume

Mercury 6.084×10^{10} km^3
5.6% that of Earth

Venus 9.284×10^{11} km^3
85.4% that of Earth

Earth 1.087×10^{12} km^3

Mars 1.643×10^{11} km^3
15.1% that of Earth

Jupiter 1.377×10^{15} km^3
1266 times that of Earth

If Jupiter were spherical, its volume would be 1.531×10^{15} km^3, which would be 1408 times that of Earth (using the equatorial diameter)

Saturn 8.183×10^{14} km^3
752 times that of Earth

If Saturn were spherical, its volume would be 9.171×10^{14} km^3, which would be 844 times that of Earth (using the equatorial diameter)

Uranus 6.995×10^{13} km^3
64.4 times that of Earth

Neptune 6.379×10^{13} km^3
58.7 times that of Earth

Pluto 6.545×10^{9} km^3
0.602% that of Earth

Wind speeds

Mercury   Mercury has such a thin atmosphere that no winds (in the traditional sense) are possible.

Venus   350 km/hr. This wind speed was measured at the tops of the Venusian clouds. Near the surface, wind speeds are very low, in the range of 0.3 to 1.0 m/s.

Earth   Earth has extremely variable wind speeds. At a minimum, there can be almost no wind (some air motion is inevitable) and at a maximum, within the vortex of a large tornado, wind speeds can reach 1739 m/s.

Mars   The Viking 1 lander measured wind speeds at 2–7 m/s in the Martian summer and 5–10 m/s in the autumn. Wind speeds during dust storms were measured at 17–30 m/s.

Jupiter   Near the Jovian equator (latitude $+30°$ to $-30°$), the wind speeds may reach 150 m/s. At latitudes further than $30°$ from the equator, maximum wind speeds are 40 m/s.

Saturn   Near the Saturnian equator (latitude $+30°$ to $-30°$), the wind speeds may reach 400 m/s. At latitudes further than $30°$ from the equator, maximum wind speeds are 150 m/s.

Uranus   Wind speeds in the Uranian atmosphere are highly variable. Velocities from 0–200 m/s have been recorded.

Neptune   Neptune has the highest wind speeds in the solar system. Velocities of 667 m/s have been recorded.

Pluto   For much of its orbit, Pluto is too far from the Sun (and thus too cold) to maintain a gaseous atmosphere. Near perihelion, some atmosphere may develop due to solar heating, but the measurement of wind speeds on Pluto (if any) remains beyond current technology.

Satellites

Note: Satellites are listed in order of increasing distance from their primary.

Albedo

		Albedo[a,b]
Earth	Moon	0.12
Mars	Phobos	0.06
	Deimos	0.07
Jupiter	Metis	0.05
	Adrastea	0.05
	Amalthea	0.05
	Thebe	0.05
	Io	0.61
	Europa	0.64
	Ganymede	0.43
	Callisto	0.20
	Leda	—
	Himalia	—
	Lysithea	—
	Elara	—
	Ananke	—
	Carme	—
	Pasiphae	—
	Sinope	—
Saturn	Pan	0.5
	Atlas	0.9
	Prometheus	0.6
	Pandora	0.9
	Hyperion	0.19–0.25
	Epimetheus	0.8
	Mimas	0.5
	Enceladus	1.0
	Calypso	0.6
	Telesto	0.5
	Tethys	0.9
	Dione	0.7
	Helene	0.7
	Rhea	0.7
	Titan	0.21
	Janus	0.8
	Iapetus	0.05–0.5
	Phoebe	0.06

		Albedo[a,b]
Uranus	Cordelia	0.07
	Ophelia	0.07
	Bianca	0.07
	Cressida	0.07
	Desdemona	0.07
	Juliet	0.07
	Portia	0.07
	Rosalind	0.07
	Belinda	0.07
	Puck	0.07
	Miranda	0.27
	Ariel	0.34
	Umbriel	0.18
	Titania	0.27
	Oberon	0.24
Neptune	Naiad	0.06
	Thalassa	0.06
	Despina	0.06
	Galatea	0.06
	Larissa	0.06
	Proteus	0.06
	Triton	0.7
	Nereid	0.2
Pluto	Charon	0.375

Notes:
[a] Values given are the most accurate available at the time of writing.
[b] Satellites with no value for their albedo have never had this quantity measured to any degree of accuracy.

Density

		Density (g/cm³)[a,b,c]
Earth	Moon	3.340
Mars	Phobos	1.750
	Deimos	1.900
Jupiter	Metis	—
	Adrastea	—
	Amalthea	—
	Thebe	—
	Io	3.57
	Europa	3.018 ± 0.035
	Ganymede	1.936 ± 0.022
	Callisto	1.851
	Leda	—
	Himalia	—
	Lysithea	—
	Elara	—
	Ananke	—
	Carme	—
	Pasiphae	—
	Sinope	—
Saturn	Pan	—
	Atlas	—
	Prometheus	0.270
	Pandora	0.420
	Hyperion	—
	Epimetheus	0.630
	Mimas	1.140
	Enceladus	1.120
	Calypso	—
	Telesto	—
	Tethys	1.000
	Dione	1.440
	Helene	—
	Rhea	1.240
	Titan	1.881
	Janus	0.650
	Iapetus	1.020
	Phoebe	—

Density (g/cm^3)[a,b,c]

Uranus	Cordelia	—
	Ophelia	—
	Bianca	—
	Cressida	—
	Desdemona	—
	Juliet	—
	Portia	—
	Rosalind	—
	Belinda	—
	Puck	—
	Miranda	1.15 ± 0.15
	Ariel	1.56 ± 0.09
	Umbriel	1.52 ± 0.11
	Titania	1.70 ± 0.05
	Oberon	1.64 ± 0.06
Neptune	Naiad	—
	Thalassa	—
	Despina	—
	Galatea	—
	Larissa	—
	Proteus	—
	Triton	2.054
	Nereid	—
Pluto	Charon	1.800

Notes:
[a] Values given are the most accurate available at the time of writing.
[b] Satellites with no value for their density have never had this quantity measured to any degree of accuracy.
[c] In order to indicate the degree of accuracy, an error range has been included where a final value of a satellite's density has not been determined.

Discoverers and dates of discovery

		Discoverer	Date
Earth	Moon	—	—
Mars	Phobos	Asaph Hall	1877 (17 Aug)
	Deimos	Asaph Hall	1877 (11 Aug)
Jupiter	Metis	Stephen P. Synnott	1979
	Adrastea	D. C. Jewitt	1979
	Amalthea	Edward Emerson Barnard	1892 (9 Sep)
	Thebe	Stephen P. Synnott	1979
	Io	Galileo Galilei	1610 (7 Jan)
	Europa	Galileo Galilei	1610 (7 Jan)
	Ganymede	Galileo Galilei	1610 (7 Jan)
	Callisto	Galileo Galilei	1610 (7 Jan)
	Leda	Charles T. Kowal	1974
	Himalia	Charles Dillon Perrine	1904
	Lysithea	Seth Barnes Nicholson	1938
	Elara	Charles Dillon Perrine	1905
	Ananke	Seth Barnes Nicholson	1951
	Carme	Seth Barnes Nicholson	1938
	Pasiphae	P. Mellote	1908
	Sinope	Seth Barnes Nicholson	1914
Saturn	Pan	M. R. Showalter	1985
	Atlas	R. Terrile	1980
	Prometheus	S. Collins and others	1980
	Pandora	S. Collins and others	1980
	Hyperion	William Lassell	1848
	Epimetheus	R. Walker and others	1980
	Mimas	Sir William Herschel	1789
	Enceladus	Sir William Herschel	1789
	Calypso	Brad Smith and others	1980
	Telesto	Brad Smith and others	1980
	Tethys	Jean-Dominique Cassini	1684
	Dione	Jean-Dominique Cassini	1684
	Helene	P. Laques and J. Lecacheux	1980
	Rhea	Jean-Dominique Cassini	1672
	Titan	Christiaan Huygens	1655
	Janus	Audouin Dollfus	1966
	Iapetus	Jean-Dominique Cassini	1671
	Phoebe	William Henry Pickering	1898

		Discoverer	Date
Uranus	Cordelia	Voyager 2[a]	1986 (20 Jan)
	Ophelia	Voyager 2[a]	1986 (20 Jan)
	Bianca	Voyager 2[a]	1986 (21 Jan)
	Cressida	Voyager 2[a]	1986 (13 Jan)
	Desdemona	Voyager 2[a]	1986 (13 Jan)
	Juliet	Voyager 2[a]	1986 (9 Jan)
	Portia	Voyager 2[a]	1986 (3 Jan)
	Rosalind	Voyager 2[a]	1986 (3 Jan)
	Belinda	Voyager 2[a]	1986 (13 Jan)
	Puck	Stephen P. Synnott	1985 (30 Dec)
	Miranda	Gerard Kuiper	1948 (16 Feb)
	Ariel	William Lassell	1851 (24 Oct)
	Umbriel	William Lassell	1851 (24 Oct)
	Titania	Sir William Herschel	1787 (11 Jan)
	Oberon	Sir William Herschel	1787 (11 Jan)
Neptune	Naiad	Voyager 2[a]	1989
	Thalassa	Voyager 2[a]	1989
	Despina	Voyager 2[a]	1989
	Galatea	Voyager 2[a]	1989
	Larissa	Stephen P. Synnott	1989
	Proteus	Stephen P. Synnott	1989
	Triton	William Lassell	1846 (10 Oct)
	Nereid	Gerard Kuiper	1949
Pluto	Charon	James Christy	1978 (22 Jun)

Note:

[a] Discovered by scientists working on the Voyager 2 mission.

Distance from planet

		Distance from planet (km)[a]
Earth	Moon	3.844×10^5
Mars	Phobos	9.377×10^3
	Deimos	2.3436×10^4
Jupiter	Metis	1.2796×10^5
	Adrastea	1.2898×10^5
	Amalthea	1.813×10^5
	Thebe	2.219×10^5
	Io	4.216×10^5
	Europa	6.709×10^5
	Ganymede	1.07×10^6
	Callisto	1.883×10^6
	Leda	1.1094×10^7
	Himalia	1.148×10^7
	Lysithea	1.172×10^7
	Elara	1.1737×10^7
	Ananke	2.12×10^7
	Carme	2.26×10^7
	Pasiphae	2.35×10^7
	Sinope	2.37×10^7
Saturn	Pan	1.33583×10^5
	Atlas	1.3764×10^5
	Prometheus	1.3935×10^5
	Pandora	1.417×10^5
	Hyperion	1.4811×10^5
	Epimetheus	1.51422×10^5
	Mimas	1.8552×10^5
	Enceladus	2.3802×10^5
	Calypso[b]	2.9466×10^5
	Telesto[b]	2.9466×10^5
	Tethys[b]	2.9466×10^5
	Dione	3.774×10^5
	Helene	3.774×10^5
	Rhea	5.2704×10^5
	Titan	1.22185×10^6
	Janus	1.51472×10^6
	Iapetus	3.5613×10^6
	Phoebe	1.2952×10^7

		Distance from planet (km)[a]
Uranus	Cordelia	4.9752×10^4
	Ophelia	5.3764×10^4
	Bianca	5.9165×10^4
	Cressida	6.1777×10^4
	Desdemona	6.2659×10^4
	Juliet	6.4358×10^4
	Portia	6.6097×10^4
	Rosalind	6.9927×10^4
	Belinda	7.5255×10^4
	Puck	8.6004×10^4
	Miranda	1.298×10^5
	Ariel	1.912×10^5
	Umbriel	2.66×10^5
	Titania	4.358×10^5
	Oberon	5.826×10^5
Neptune	Naiad	4.8227×10^4
	Thalassa	5.0075×10^4
	Despina	5.2526×10^4
	Galatea	6.1953×10^4
	Larissa	7.3548×10^4
	Proteus	$1.176\,47 \times 10^5$
	Triton	3.5476×10^5
	Nereid[c]	5.5134×10^6
Pluto	Charon	1.9405×10^4

Notes:

[a] Values given are the most accurate available at the time of writing.

[b] It is interesting that Calypso, Telesto and Tethys all lie the same distance from Saturn.

[c] Average value.

Eccentricity

		Eccentricity[a,b]
Earth	Moon	0.0549
Mars	Phobos	0.0151
	Deimos	0.00033
Jupiter	Metis	0.041
	Adrastea	~0
	Amalthea	0.003
	Thebe	0.0015
	Io	0.041
	Europa	0.0101
	Ganymede	0.0015
	Callisto	0.007
	Leda	0.163
	Himalia	0.163
	Lysithea	0.107
	Elara	0.207
	Ananke	0.169
	Carme	0.207
	Pasiphae	0.378
	Sinope	0.275
Saturn	Pan	~0
	Atlas	~0
	Prometheus	0.0042
	Pandora	0.0042
	Hyperion	0.1042
	Epimetheus	0.009
	Mimas	0.0202
	Enceladus	0.0045
	Calypso	~0
	Telesto	~0
	Tethys	0
	Dione	0.0022
	Helene	0.005
	Rhea	0.001
	Titan	0.0292
	Janus	0.007
	Iapetus	0.0283
	Phoebe	0.163

71

Eccentricity[a,b]

Uranus	Cordelia	$0.000\,469 \pm 0.000\,41$
	Ophelia	$0.010\,14 \ \ \pm 0.000\,404$
	Bianca	$0.000\,878 \pm 0.000\,518$
	Cressida	$0.000\,091 \pm 0.000\,349$
	Desdemona	$0.000\,227 \pm 0.000\,297$
	Juliet	$0.000\,233 \pm 0.000\,321$
	Portia	$0.000\,165 \pm 0.000\,365$
	Rosalind	$0.000\,585 \pm 0.000\,249$
	Belinda	$0.000\,109 \pm 0.000\,206$
	Puck	$0.000\,051 \pm 0.000\,178$
	Miranda	0.0027
	Ariel	0.0034
	Umbriel	0.0050
	Titania	0.0022
	Oberon	0.0008
Neptune	Naiad	0.00
	Thalassa	0.00
	Despina	0.00
	Galatea	0.00
	Larissa	0.00
	Proteus	0.00
	Triton	0.00
	Nereid	0.7512
Pluto	Charon	0.00

Notes:
[a] Values given are the most accurate available at the time of writing.
[b] In order to indicate the degree of accuracy, an error range has been included where the final value of a satellite's eccentricity has not been determined.

Inclination of orbit

		Inclination[a,b]
Earth	Moon	5.145°
Mars	Phobos	1.08°
	Deimos	1.79°
Jupiter	Metis	~0°
	Adrastea	~0°
	Amalthea	0.40°
	Thebe	0.8°
	Io	0.040°
	Europa	0.470°
	Ganymede	0.195°
	Callisto	0.281°
	Leda	27°
	Himalia	0.28°
	Lysithea	29°
	Elara	28°
	Ananke	147°
	Carme	163°
	Pasiphae	148°
	Sinope	153°
Saturn	Pan	~0°
	Atlas	~0°
	Prometheus	0.0°
	Pandora	0.0°
	Hyperion	0.43°
	Epimetheus	0.34°
	Mimas	1.53°
	Enceladus	0.02°
	Calypso	~0°
	Telesto	~0°
	Tethys	1.09°
	Dione	0.02°
	Helene	0.2°
	Rhea	0.35°
	Titan	0.33°
	Janus	0.14°
	Iapetus	7.52°
	Phoebe	175.3°

		Inclination[a,b]
Uranus	Cordelia	$0.140° \pm 0.098°$
	Ophelia	$0.091° \pm 0.272°$
	Bianca	$0.156° \pm 0.172°$
	Cressida	$0.282° \pm 0.116°$
	Desdemona	$0.160° \pm 0.117°$
	Juliet	$0.042° \pm 0.140°$
	Portia	$0.087° \pm 0.151°$
	Rosalind	$0.057° \pm 0.113°$
	Belinda	$0.033° \pm 0.108°$
	Puck	$0.314° \pm 0.079°$
	Miranda	$4.22°$
	Ariel	$0.31°$
	Umbriel	$0.36°$
	Titania	$0.10°$
	Oberon	$0.10°$
Neptune	Naiad	$4.74°$
	Thalassa	$0.21°$
	Despina	$0.07°$
	Galatea	$0.05°$
	Larissa	$0.20°$
	Proteus	$0.55°$
	Triton	$156.834°$
	Nereid	$7.23°$
Pluto	Charon	$96.56°$

Notes:
[a] Values given are the most accurate available at the time of writing.
[b] In order to indicate the degree of accuracy, an error range has been included where the final value of a satellite's inclination has not been determined.

Mass

		Mass (kg)[a,b]
Earth	Moon	7.15×10^{22}
Mars	Phobos	1.8×10^{15}
	Deimos	1.08×10^{16}
Jupiter	Metis	—
	Adrastea	—
	Amalthea	—
	Thebe	—
	Io	8.94×10^{22}
	Europa	4.799×10^{22}
	Ganymede	1.482×10^{23}
	Callisto	1.076×10^{23}
	Leda	—
	Himalia	—
	Lysithea	—
	Elara	—
	Ananke	—
	Carme	—
	Pasiphae	—
	Sinope	—
Saturn	Pan	—
	Atlas	—
	Prometheus	1.4×10^{17}
	Pandora	1.3×10^{17}
	Hyperion	—
	Epimetheus	5.05×10^{17}
	Mimas	3.75×10^{19}
	Enceladus	7.3×10^{19}
	Calypso	—
	Telesto	—
	Tethys	6.22×10^{20}
	Dione	1.052×10^{21}
	Helene	—
	Rhea	2.31×10^{21}
	Titan	1.3455×10^{23}
	Janus	4.98×10^{18}
	Iapetus	1.59×10^{21}
	Phoebe	—

Mass (kg)[a,b]

Uranus	Cordelia	—
	Ophelia	—
	Bianca	—
	Cressida	—
	Desdemona	—
	Juliet	—
	Portia	—
	Rosalind	—
	Belinda	—
	Puck	—
	Miranda	6.59×10^{19}
	Ariel	1.353×10^{21}
	Umbriel	1.172×10^{21}
	Titania	3.527×10^{21}
	Oberon	3.014×10^{21}
Neptune	Naiad	—
	Thalassa	—
	Despina	—
	Galatea	—
	Larissa	—
	Proteus	—
	Triton	2.147×10^{22}
	Nereid	—
Pluto	Charon	1.7×10^{21}

Notes:
[a] Values given are the most accurate available at the time of writing.
[b] Satellites with no value for their mass have never had this quantity measured to any degree of accuracy.

Orbital period

Orbital period[a]

		Days	Days–hours–minutes
Earth	Moon	27.322	27^{d}07^{h}43.7^m
Mars	Phobos	0.31891	07^{h}39.2^m
	Deimos	1.26244	01^{d}06^{h}17.9^m
Jupiter	Metis	0.29478	07^{h}04.5^m
	Adrastea	0.29826	07^{h}09.5^m
	Amalthea	0.498179	11^{h}57.4^m
	Thebe	0.6745	16^{h}11.3^m
	Io	1.769167	01^{d}18^{h}27.6^m
	Europa	3.551810	03^{d}13^{h}14.6^m
	Ganymede	7.154553	07^{d}03^{h}42.6^m
	Callisto	16.689018	16^{d}16^{h}32.2^m
	Leda	238.72	238^{d}17^{h}16.8^m
	Himalia	250.5662	250^{d}13^{h}35.3^m
	Lysithea	259.22	259^{d}05^{h}16.8^m
	Elara	259.6528	259^{d}15^{h}40^m
	Ananke	631 retrograde	631^d (r)
	Carme	692 retrograde	692^d (r)
	Pasiphae	735 retrograde	735^d (r)
	Sinope	758 retrograde	758^d (r)
Saturn	Pan	0.575	13^{h}48^m
	Atlas	0.6019	14^{h}26.7^m
	Prometheus	0.612986	14^{h}42.7^m
	Pandora	0.628804	15^{h}05.5^m
	Hyperion	21.276609	21^{d}06^{h}38.3^m
	Epimetheus	0.69459	16^{h}40.2^m
	Mimas	0.9424218	22^{h}37.1^m
	Enceladus	1.370218	01^{d}08^{h}53.1^m
	Calypso	1.887802	01^{d}21^{h}18.4^m
	Telesto	1.887802	01^{d}21^{h}18.4^m
	Tethys	1.887802	01^{d}21^{h}18.4^m
	Dione	2.736915	02^{d}17^{h}41.2^m
	Helene	2.736915	02^{d}17^{h}41.2^m
	Rhea	4.5175	04^{d}12^{h}25.2^m
	Titan	21.276609	21^{d}06^{h}38.3^m
	Janus	0.69459	16^{h}40.2^m
	Iapetus	79.330183	79^{d}07^{h}55.5^m
	Phoebe	550.48 retrograde	550^{d}11^{h}31.2^m (r)

Orbital period[a]

		Days	Days–hours–minutes
Uranus	Cordelia	0.335033	08^{h}02.4^m
	Ophelia	0.376409	09^{h}02^m
	Bianca	0.434577	10^{h}25.8^m
	Cressida	0.46357	11^{h}07.5^m
	Desdemona	0.473651	11^{h}22.1^m
	Juliet	0.493066	11^{h}50^m
	Portia	0.513196	12^{h}19^m
	Rosalind	0.558459	13^{h}24.2^m
	Belinda	0.623525	14^{h}57.9^m
	Puck	0.761832	18^{h}17^m
	Miranda	1.413	01^{d}09^{h}54.7^m
	Ariel	2.52	02^{d}12^{h}28.8^m
	Umbriel	4.144	04^{d}03^{h}27.4^m
	Titania	8.706	08^{d}16^{h}56.6^m
	Oberon	13.463	13^{d}11^{h}06.7^m
Neptune	Naiad	0.294396	07^{h}03.9^m
	Thalassa	0.311485	07^{h}28.5^m
	Despina	0.334655	08^{h}01.9^m
	Galatea	0.428745	10^{h}17.4^m
	Larissa	0.554654	13^{h}18.7^m
	Proteus	1.122315	01^{d}02^{h}56.1^m
	Triton	5.876854 retrograde	05^{d}21^{h}02.7^m (r)
	Nereid	360.136 19	360^{d}03^{h}16.11^m
Pluto	Charon	6.387	06^{d}09^{h}17.3^m

Note:

[a] Values given are the most accurate available at the time of writing.

Size

		Equatorial diameter (km)[a]	Ratio: satellite/planet
Earth	Moon	3476	0.2764
Mars	Phobos	26(×18)	0.0038
	Deimos	16(×10)	0.0024
Jupiter	Metis	50	0.0003
	Adrastea	30	0.0002
	Amalthea	262(×146×134)	0.0018
	Thebe	110	0.0008
	Io	3630	0.0254
	Europa	3120	0.0218
	Ganymede	5268	0.0369
	Callisto	4806	0.0336
	Leda	8	0.0001
	Himalia	180	0.0013
	Lysithea	24	0.0002
	Elara	90	0.0006
	Ananke	20	0.0001
	Carme	30	0.0002
	Pasiphae	36	0.0003
	Sinope	28	0.0002
Saturn	Pan	20	0.0002
	Atlas	37(×34.4×27)	0.0003
	Prometheus	1528	0.0127
	Pandora	110(×88×62)	0.0009
	Hyperion	370(×280×226)	0.0031
	Epimetheus	138(×110×110)	0.0011
	Mimas	398	0.0033
	Enceladus	498	0.0041
	Calypso	30(×16×16)	0.0002
	Telesto	30(×25×15)	0.0002
	Tethys	1060	0.0088
	Dione	1120	0.0093
	Helene	32	0.0003
	Rhea	1530	0.0127
	Titan	5150	0.0427
	Janus	199(×191×151)	0.0017
	Iapetus	1436	0.0119
	Phoebe	230(×220×210)	0.0019

		Equatorial diameter (km)[a]	Ratio: satellite/planet
Uranus	Cordelia	26	0.0005
	Ophelia	32	0.0006
	Bianca	44	0.0009
	Cressida	66	0.0013
	Desdemona	58	0.0011
	Juliet	42	0.0008
	Portia	110	0.0022
	Rosalind	58	0.0011
	Belinda	68	0.0013
	Puck	154	0.0030
	Miranda	480($\times$ 468 $\times$ 465)	0.0094
	Ariel	1162($\times$ 1156 $\times$ 1156)	0.0227
	Umbriel	1169	0.0229
	Titania	1578	0.0309
	Oberon	1523	0.0298
Neptune	Naiad	58	0.0012
	Thalassa	80	0.0016
	Despina	148	0.0030
	Galatea	158	0.0032
	Larissa	208($\times$ 178)	0.0042
	Proteus	436($\times$ 416 $\times$ 402)	0.0088
	Triton	2705	0.0546
	Nereid	340	0.0069
Pluto	Charon	1172	0.5052

Note:

[a] Many of the smaller solar system bodies are far from round. At such low mass there is not enough gravity to form a spherical shape. Collisions could also have affected the present shapes of these objects. Dimensions of these satellites (length, width, height) have been provided.

Part 2

Planets

Mercury

Physical data

Size   4879.4 km

Mass 3.303×10^{23} kg

Escape velocity   4.25 km/s (15 300 km/hr)

Temperature range

Minimum	Average	Maximum
$-173\,°C$	$179\,°C$	$427\,°C$

Oblateness*   0

Surface gravity 2.78 m/s^2

Volume 6.084×10^{10} km^3
5.6% that of Earth

Magnetic field strength and orientation   Mercury has slightly more than 1% of the magnetic field of the Earth. The fact that it has any magnetic field at all is amazing, considering the slowness of Mercury's rotation. The field is probably due to its composition. Mercury has an iron core, much like the Earth. The core of Mercury comprises approximately 70–80% of the planet's radius, the outer region being largely composed of silicate rocks.

Albedo (visual geometric albedo)   0.10

Density (water = 1)   5.43 g/cm^3

Solar irradiance   3566 watts/m^2

Atmospheric pressure   0

Composition of atmosphere

Potassium (K)	31.7%
Sodium (Na)	24.9%
Oxygen (O)	9.5%
Argon (Ar)	7.0%
Helium (He)	5.9%
Molecular oxygen (O_2)	5.6%
Molecular nitrogen (N_2)	5.2%
Carbon dioxide (CO_2)	3.6%
Water vapor (H_2O)	3.4%
Molecular hydrogen (H_2)	3.2%

Maximum wind speeds   Mercury has such a thin atmosphere that no winds (in the traditional sense) are possible.

* There is no appreciable difference between the equatorial diameter of Mercury and its polar diameter. This is almost certainly due to the slowness of the planet's rotation. Note: the term "oblateness" is sometimes referred to as a planet's "ellipticity."

Outstanding surface features

Mercury is the second smallest planet, not much larger than the Moon. Even so, 297 features have been named on the planet. Here are the largest of Mercury's features:

Feature	Diameter (km)
Antoniadi Ridge	450
Beethoven Crater	643
Caloris Basin	1287
Dostoevski Crater	411
Goethe Crater	383
Haydn Crater	270
Homer Crater	314
Monet Crater	303
Mozart Crater	270
Raphael Crater	343
Shakespeare Crater	370
Tolstoy Crater	390
Van Eyck Crater	282
Vyasa Crater	290

Orbital data

Period of rotation   58.6462 days
1407.5 hours
$58^d15^h30.5^m$

Period of revolution (sidereal orbital period)   0.24085 years
87.969 days
$0^y87^d23.3^h$

Synodic period $115^d21^h07.2^m$

Equatorial velocity of rotation   10.891 km/hr

Velocity of revolution   47.88 km/s (172 368 km/hr)

Distance from Sun   Average   0.3871 AU
57 910 000 km

Maximum   0.4667 AU
69 815 900 km

Minimum   0.3075 AU
46 003 500 km

Distance from Earth   Maximum   1.48 AU
221 920 880 km

Minimum   0.52 AU
77 269 900 km

Apparent size of Sun (average)   1.38°

Apparent brightness of Sun $m_{\text{vis}} = -28.77$

Orbital eccentricity   0.2056

Orbital inclination   7.004°

Inclination of equator to orbit   0°

Observational data

Maximum angular distance from Sun   28°

Greatest brilliancy   −1.3

Angular size*   Maximum 10″
                    Minimum   4.9″

Early ideas

The Sumerians named the planet Mercury Ubu-idim-gud-ud. The Babylonians called it gu-ad or gu-utu, and it was this early group of people who recorded the first detailed observations of the planet. There are existing tablets which show that the Babylonians were very careful observers of Mercury. They recorded six dates of importance in each of Mercury's cycles: the first visible heliacal rising of the planet; the start of its retrogression; the beginning of the period in which the planet is too near the Sun to be seen; the end of the "invisible" period; the end of retrogression; and the last visible heliacal setting of the planet.

In Egypt, the planet Mercury was identified with the god Thoth, a manifestation of the creative force of the universe. Thoth was the spouse of the goddess Maat, and he took over her powers and spoke the word of creation. To the early Chinese, the planet was known as Shui xing.

Mercury was the Roman name for the Greek god Hermes. His Latin name was apparently derived from *merx* or *mercator*, a merchant. As a god of commerce as well as magic, illusion and trickery, Mercury was often regarded as very clever but not altogether trustworthy. In both Greece and Rome he was the patron of mercantile endeavors, metals, smithcraft and the arts of natural science (that became medieval alchemy). He gave his name to the only metallic element that is liquid at ordinary temperatures, often called quicksilver, a name which literally means "living silver."

The sign of the planet Mercury also represents Wednesday, which was *dies mercurii* to the Romans. In English, this became Woden's day, due to the fact that Mercury was identified with Woden in northern Europe. Among each of the barbarian tribes Mercury was associated with one of the major gods. Like Woden, Mercury was also a god of the dead. His sign was found in catacombs and other subterranean places of worship.

For centuries, Mercury remained an enigma. It was a small planet, and visual observations were difficult due to the fact that it never ventured far

---

* This measurement is the apparent angular diameter of Mercury, measured in seconds of arc, as seen from Earth.

A very early image of Mercury. Fontana, Francesco, *Novae coelestium terrestriumq[ue] rerum observationes*, Naples, 1646. (Linda Hall Library)

beyond the region of the Sun. Observations of detail on Mercury were first announced by the German amateur astronomer Johann H. Schroeter from his private observatory at Lilienthal near Bremen. Schroeter claimed to have seen one of the horns of a crescent Mercury blunted. (He made the same observational claim for Venus.) He attributed this to the presence of a mountain (20 km in height). Sir William Herschel attempted, but was unable, to verify this sighting. From Schroeter's observations, F. W. Bessel obtained a rotational period of $24^h00^m53^s$. Bessel also calculated the tilt of Mercury's axis to be $70°$.

During the nineteenth century, a number of observers claimed to see surface detail on Mercury. The British astronomer William Frederick Denning made a series of sketches in 1881, and obtained a rotational period of 25 hours. In 1892, the French astronomer Léopold Trouvelot seemed to find the same features as Schroeter had, but his observations were never confirmed.

It was not until the Italian astronomer Giovanni Virginio Schiaparelli made a detailed map of Mercury in 1889, that all doubt as to the reality of surface features on Mercury was erased. At the Brera Observatory in Milan, Schiaparelli used 22-cm and 49-cm refractors and made his observations during daylight hours, when Mercury was high in the sky. He concluded that Mercury was tidally locked to the Sun, always keeping the same face toward it, as the Moon does to the Earth. He published a rotational period for Mercury of 88 days.

In 1896, at the Lowell Observatory, Percival Lowell started a series of observations of Mercury with the 61-cm refractor. His results agreed with Schiaparelli's with regard to the synchronous orbital period of Mercury. The map which he created was covered with dark lines and patches. Lowell explained the markings as the result of planet-wide cooling. Unfortunately, as with his observations of Mars, no other observers saw what Lowell imagined.

T. J. J. See, working with the 66-cm refractor at the US Naval Observatory, claimed to have observed a large number of craters on Mercury. In June, 1901, See made a drawing of Mercury showing, among others, a very large crater. Some modern-day astronomers have suggested that See actually observed the crater Beethoven, the largest on Mercury. Other astronomers point out that at the time Mercury's angular diameter was only 6.6 seconds of arc, and that such an observation would be impossible.

Perhaps the greatest of the twentieth century observers of Mercury was the Greek-born French astronomer Eugène Marie Antoniadi. His numerous observations of Mercury were always made during daylight hours, just like Schiaparelli. Antoniadi published *La Planète Mercure* in 1934, an in-depth look at the planet. The book contained maps of various features that Antoniadi claimed to have seen. However, as with those of See, the observations of Antoniadi have been questioned by modern astronomers.

Important concepts

The interior composition of Mercury

On 29 March 1974, the conceptions of planetary scientists who study Mercury changed. It was on this date that Mariner 10 made the first of three close encounters with Mercury. Among the wealth of data returned to Earth was a surprise – Mercury has an Earth-like magnetic field strong enough to deflect the solar wind. The magnetometer instrumentation on Mariner 10 picked up the field on the first and the third of three flybys of the planet. No data regarding Mercury's magnetosphere was obtained from the second encounter

Drawings of Mercury. Villiger, Walter, "Die Rotationszeit des Planeten Venus ..." *Neue Annalen der K. Sternwarte in München,* **3**:301–42, Munich, 1898. (Linda Hall Library)

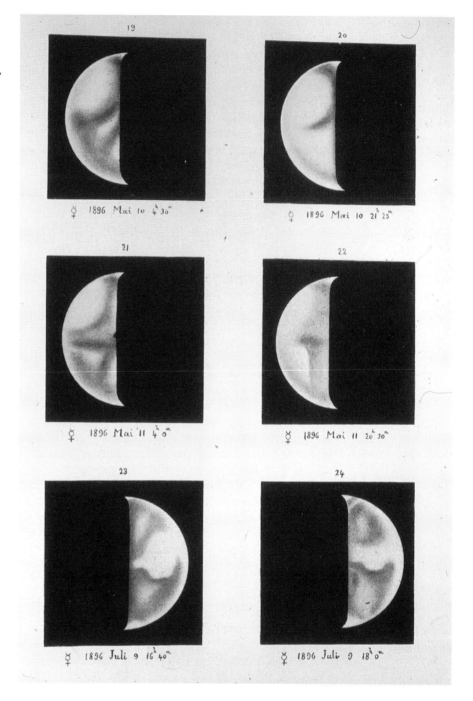

Drawings of round balls set up at a distance and telescopically observed. This image says a great deal about perception. Villiger, Walter, "Die Rotationszeit des Planeten Venus …" *Neue Annalen der K. Sternwarte in München,* **3**:301–42, Munich, 1898. (Linda Hall Library)

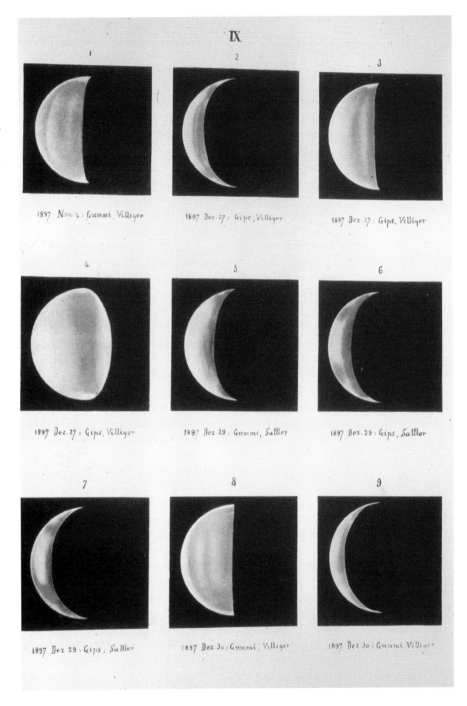

because the trajectory of the spacecraft was altered to permit optimal imaging of the sunlit side of the planet. The closest approach distance of 50 000 km was outside the limit of Mercury's magnetosphere.

What is responsible for the magnetic field? After the initial Mariner 10 encounter, it was postulated that the observed magnetic field might be due to a solar wind induction effect. This was proven incorrect through measurements taken during the third spacecraft encounter. The strength of the Mercurian field is approximately 20 times greater than the field measured in the interplanetary medium. Thus, the field is not induced by the Sun, but rather is internal to the planet.

Two possibilities remain. First, the magnetic field may be generated by a dynamo-like effect within the core of the planet, such as is the case with the Earth. The other possibility is referred to as "remnant magnetization." This theory states that, during its formation, Mercury possessed a magnetic field due to a molten core (the dynamo effect). As the planet cooled and the core solidified, the magnetic field should have decreased to zero. It did not, however, because the material of which Mercury is made retained some of the original magnetic characteristics induced by the early dynamo – in essence, the planet was magnetized.

Mercury is the planet with the highest uncompressed density, 5.3 g/cm^3, as opposed to Earth's 4.45 g/cm^3. Planetary scientists have calculated that to have such a high average density, Mercury's composition must be approximately 70% metallic and 30% silicate. Estimates which take into account this model of the interior of Mercury give a temperature at the boundary of the core and mantle too low to maintain an iron–nickel core in a molten state. If this is true, some other process must be at work. It has been suggested that the core of Mercury may contain a significant amount of material which has a lower melting point. The substance most often mentioned is sulfur.

Several models of the interior of Mercury have been proposed. The model preferred by planetologists assigns 88–91% of the core to iron, 6.5–7.5% to nickel and 0.5–5% to sulfur. Alternatively, a volatile-rich model has been proposed, containing a core composed of 76.22% iron, 6.2% nickel and 17.58% sulfur. Such a model is consistent with the observed average density of the planet. Astronomers are unsure how Mercury arrived at its current high density. No model of solar system formation can account for this through normal processes of accretion and condensation.

One plausible theory states that a giant impacting body collided with Mercury soon after its formation. In this model, the original mass of Mercury would have been approximately 2.25 times its current mass. This is assumed so as to give the "original" Mercury a silicate-to-iron ratio of common chondritic material. The impactor has been given a mass of approximately one-sixth of the above value. It is theorized that much of the crust and upper mantle would have been stripped away by such an event. The material lost would contain high percentages of calcium and aluminum, accounting for their present low percentages. Such a collision is now widely accepted as the mechanism by which the Moon formed.

Yet another speculation assumes that Mercury formed very early in the history of the solar system, before the Sun had stabilized. This proto-Mercury would have been slightly more than twice the mass of the current planet. As the proto-Sun began to contract, temperatures in the vicinity of Mercury would be between 2500 K and 3500 K and may, for a time, have reached tens of thousands of degrees. Much of the surface rock of Mercury would have vaporized at these temperatures, forming an atmosphere composed of rock vapor. The pressure of the nebular wind would have carried the atmosphere away from Mercury, leaving the metal-rich body we now observe.

The atmosphere of Mercury

It is a commonly held belief that Mercury has no atmosphere. It is a generalization which, though incorrect, is not too far from fact. The density of the atmosphere is so low that the atmosphere/space boundary is usually considered to be the surface of the planet. In fact, gaseous atoms collide with Mercury's surface more often than with each other.

In 1974, using its airglow (ultraviolet) spectrometer, the Mariner 10 spacecraft established the presence of hydrogen, helium and oxygen. In 1985, ground-based observation revealed sodium atoms and in 1986, potassium was observed. Both of these atmospheric elements were discovered by spectroscopically analyzing the scattering of sunlight.

The abundances of atmospheric elements on Mercury are as follows:

Element	Symbol	Abundance per cm^3
Potassium	K	1.4×10^8
Sodium	Na	1.1×10^8
Oxygen	O	4.2×10^7
Argon	Ar	3.1×10^7
Helium	He	2.6×10^7
Molecular oxygen	O_2	2.5×10^7
Molecular nitrogen	N_2	2.3×10^7
Carbon dioxide	CO_2	1.6×10^7
Water	H_2O	1.5×10^7
Molecular hydrogen	H_2	1.4×10^7

It is interesting to note that the abundance of helium changes dramatically as the Mercurian day becomes night. The proportion of helium between the point directly under the Sun when it is on the meridian (the subsolar point) and the mid-night point directly opposite (the antisolar point) differs by a factor of at least 50, and possibly as much as 150. As one might imagine, when the Sun is shining there are a lot more processes occurring which can decrease the atmospheric density of helium.

The individual components of the Mercurian atmosphere – that is to say, the atoms themselves – are short-lived with respect to their connection to Mercury. A number of processes cause the loss of these atoms into space, the most predominant being photoionization. During the Mercurian day, the lifetime of a sodium or potassium atom is approximately 3 hours. This lifetime is reduced by 50% at perihelion. Whether or not an ion is lost from the planet depends on several external factors, among them the strength of the solar wind and the properties of Mercury's magnetosphere at that moment.

If the loss of atoms into space were the only factor, then assuredly Mercury would have no atmosphere, and this would occur in a very short time. There are, however, several mechanisms which are replacing the scant atmosphere. One such process involves the capture of ions from the solar wind. Mercury's magnetosphere is just strong enough to deflect some of the incoming atoms toward its surface, where they impact and are trapped.

A process which may help to explain some of the sodium and potassium observed in the thin Mercurian atmosphere is impact by meteoroids. Even a relatively slow-moving meteor (a few kilometers per second) would create some vapor when it struck the surface. For bodies traveling at faster rates (30 km/s) the vapor production would be many times that of the impactor's mass.

Another contributor to the atmosphere of Mercury is the direct thermal evaporation of atoms. Also, atoms may be sputtered off the surface of Mercury. This process implies contact with an incoming atom or ion, but does not involve ionization of the escaping atom.

Proving the theory of relativity by calculating Mercury's orbital precession

Since the moment the British physicist Sir Isaac Newton proposed the Law of Universal Gravitation in the seventeenth century, the science of astronomy was given a predictive tool like none before or since. The positions of the planets and their satellites could be predicted with great accuracy, the motions of comets were no longer a surprise, and even unseen worlds like Neptune could be discovered. It seemed that the great Newton had bequeathed to science a tool without limits. Well, almost.

In the nineteenth century, careful observations revealed that the perihelion point of the orbit of Mercury was changing, advancing, in fact. Gravitational theorists set to work. Venus, it was shown, caused Mercury's perihelion to advance by 277 seconds of arc per century. This amount, less than a tenth of a degree in one hundred years, is significant when viewed with reference to astronomical time scales. Furthermore, Venus was not the sole influence. The Earth added 91 seconds of arc to Mercury's perihelion advance and Mars an additional 8 seconds of arc. Jupiter, although much more distant, is the second most massive body in the solar system. Its contribution amounted to 153 seconds of arc. The remaining planets add approximately 2 seconds of arc, bringing the total to 531 seconds of arc – according to Newton and the law of gravity.

When the actual measurements were done, it was seen that the perihelion point of Mercury's orbit actually advances by 574 seconds of arc per century.

Sir Isaac Newton
as pictured in the
3rd edition of the
*Principia
Mathematica.*
(Linda Hall Library)

The difference in these two values – 43 seconds of arc – was one of the great mysteries of celestial mechanics for more than half a century. Many ideas were proposed to account for this discrepancy.

The French astronomer Urbain Jean Joseph LeVerrier proposed that there was an additional planet lying at a distance of 30 000 000 km from the Sun. (See the next section for more information on this proposed planet.)

Another theory proposed a large quantity of fine dust in the region between the Sun and Mercury. Many years of careful observation turned up no evidence whatsoever for such an area. Finally, in 1916, the German astronomer Karl Schwarzschild, using the General Theory of Relativity developed by Albert Einstein, accurately accounted for the precession of the perihelion of Mercury.

According to the theory, Mercury's motion produces both momentum and energy (including gravitational energy) that, when viewed relativistically, add to the overall mass of the planet. If Mercury is more massive than can be accounted for by Newtonian mechanics, then the Law of Universal Gravitation will not be able to account for the motion of the planet. This was especially noticeable with regard to the perihelion point of Mercury. Mercury has a very eccentric orbit (0.2056). When Mercury is nearest the Sun, its orbital velocity is at a maximum, and so is the gravitational force exerted by the Sun. The addition of the relativistic mass to the overall mass of Mercury produces a very small acceleration to the orbital motion of the planet. The General Theory of Relativity correctly predicted an increase in the orbital perihelion point equivalent to an additional 43 seconds of arc per century. This means that Mercury's perihelion point will make one complete circuit and return to its present position after approximately 30140 years.

Theories of intra-Mercurial planets

During the nineteenth century, as celestial mechanics was being refined, it was noticed that Mercury did not move exactly as predicted. In a lecture on 2 Jan 1860, the French astronomer and mathematician Urbain LeVerrier announced that the problem of observed deviations of the motion of Mercury could be solved by assuming an intra-Mercurial planet, or possibly a second asteroid belt inside Mercury's orbit. This planet would be too close to the Sun for normal observations. However, LeVerrier proposed that the time to observe this intra-Mercurial planet (or belt of asteroids) was during a transit or a total solar eclipse. Almost immediately, reports started arriving. Suspicious dots and spots and disks were seen or imagined, and orbits were calculated for each object.

Several months earlier, LeVerrier had received a letter from the amateur astronomer E. Lescarbault, who reported having seen a round black spot on the Sun on 26 Mar 1859. Lescarbault assumed that it was a planet transiting the Sun. LeVerrier investigated this observation and soon had an orbit computed. The calculated diameter was considerably smaller than Mercury's and its mass was estimated at 0.059 of Mercury's mass. If this were a planet, its size was too small to account for the deviations of Mercury's orbit. LeVerrier considered another option: perhaps this was the largest member of the intra-Mercurial asteroid belt. LeVerrier was convinced of the reality of the object and named it Vulcan.

Solar eclipses always generate a great deal of activity. In 1860, however, the excitement was heightened by the distinct possibility that a new planet could be discovered. LeVerrier communicated all observations and the orbit of the possible object and then mobilized several French eclipse teams and a few others for the express purpose of locating Vulcan. Unfortunately, the "planet" went unobserved.

Title page to Isaac
Newton's *Principia
Mathematica*.
(Linda Hall Library)

PHILOSOPHIÆ

NATURALIS

PRINCIPIA

MATHEMATICA.

Autore *JS. NEWTON*, *Trin. Coll. Cantab. Soc.* Matheseos
Professore *Lucasiano*, & Societatis Regalis Sodali.

IMPRIMATUR·

S. PEPYS, *Reg. Soc.* PRÆSES.

Julii 5. 1686:

LONDINI,

Jussu *Societatis Regiæ* ac Typis *Josephi Streater*. Prostant Vena-
les apud *Sam. Smith* ad insignia Principis *Walliæ* in Cœmiterio
D. *Pauli*, aliosq; nonnullos Bibliopolas. *Anno* MDCLXXXVII.

During the total solar eclipse of 29 Jul 1878, two astronomers claimed to have seen in the vicinity of the Sun small illuminated disks which could only be small planets inside Mercury's orbit. J. C. Watson (professor of astronomy at the University of Michigan) observed what he listed as two intra-Mercurial planets. He surmised that one of these was surely Vulcan. Lewis Swift (co-discoverer of Comet Swift–Tuttle), also saw an object he believed to be Vulcan. Amazingly, its position was different than either of Watson's two "intra-Mercurials." Orbits were quickly calculated for these objects. However, neither Watson's nor Swift's observations could be reconciled with LeVerrier's Vulcan.

These were the last reported observations of Vulcan, in spite of several searches at different total solar eclipses. In 1916, Albert Einstein published his General Theory of Relativity, which explained the deviations in the motions of Mercury without the need to invoke an unknown intra-Mercurial planet.

It is possible that Lescarbault happened to see a small asteroid passing very close to the Earth, just inside Earth's orbit. Asteroids lying outside the traditional belt between Mars and Jupiter were unknown at that time, so Lescarbault and LeVerrier naturally assumed that the observations were of an intra-Mercurial planet.

Interesting facts

It is conceivable that, at a certain point on Mercury, a theoretical observer could see the Sun rise twice in the same Mercurian day. This can occur because approximately four days prior to perihelion, Mercury's orbital speed exactly equals its rotational rate. At this point, the Sun's normal east-to-west motion first becomes stationary, and then the Sun reverses direction. While passing perihelion, Mercury revolves faster than it rotates. The solar retrograde motion continues until approximately four days after perihelion, when the direct motion once again takes over. The theoretical observer would see the Sun rise approximately halfway, then set completely, then rise again and move normally across the sky!

Mercury has a total of 297 named features, of which 239 are craters.

The largest crater on Mercury is Beethoven, with a diameter of 643 km. It is the largest crater in the solar system.

From Mercury, the Sun is 6.3 times brighter than from Earth.

For telescopic observers, the easiest "marking" which may be observed on Mercury has been described as a slight blunting of the planet's south cusp, presumably due to a darker surface at high southern latitudes.

Mercury's rotational rate of 58.6461 ± 0.005 days is essentially 2/3 of its orbital period of 87.9693 days.

The light from the Sun takes, on average, 3 minutes 13 seconds to reach Mercury. At Mercury's perihelion, this time is reduced to 2 minutes 33 seconds, and at aphelion it is 3 minutes 53 seconds.

Observing data   *Future dates of conjunction*

Dates of conjunction	
Inferior	Superior
26 Jul 1999	8 Sep 1999
15 Nov 1999 (transit)	16 Jan 2000
1 Mar 2000	9 May 2000 (transit)
6 Jul 2000	22 Aug 2000
30 Oct 2000	25 Dec 2000
13 Feb 2001	23 Apr 2001
16 Jun 2001	5 Aug 2001
14 Oct 2001	4 Dec 2001
27 Jan 2002	7 Apr 2002
27 May 2002	21 Jul 2002
27 Sep 2002	14 Nov 2002 (transit)
11 Jan 2003	21 Mar 2003
7 May 2003 (transit)	5 Jul 2003
11 Sep 2003	25 Oct 2003
27 Dec 2003	4 Mar 2004
17 Apr 2004	18 Jun 2004
23 Aug 2004	5 Oct 2004
10 Dec 2004	14 Feb 2005
29 Mar 2005	3 Jun 2005
5 Aug 2005	18 Sep 2005
24 Nov 2005	27 Jan 2006
12 Mar 2006	19 May 2006
18 Jul 20006	1 Sep 2006
9 Nov 2006	7 Jan 2007
23 Feb 2007	3 May 2007
28 Jun 2007	16 Aug 2007
24 Oct 2007	17 Dec 2007
6 Feb 2008	16 Apr 2008
7 Jun 2008	29 Jul 2008
7 Oct 2008	25 Nov 2008
20 Jan 2009	31 Mar 2009
18 May 2009	14 Jul 2009
20 Sep 2009	5 Nov 2009
4 Jan 2010	14 Mar 2010
29 Apr 2010	28 Jun 2010
3 Sep 2010	17 Oct 2010
20 Dec 2010	

Future greatest elongation dates

East of Sun (evening sky)	West of Sun (morning sky)
24 Oct 1999	14 Aug 1999
15 Feb 2000	3 Dec 1999
9 Jun 2000	28 Mar 2000
6 Oct 2000	27 Jul 2000
28 Jan 2001	15 Nov 2000
22 May 2001	11 Mar 2001
18 Sep 2001	9 Jul 2001
11 Jan 2002	29 Oct 2001
4 May 2002	21 Feb 2002
1 Sep 2002	21 Jun 2002
26 Dec 2002	13 Oct 2002
16 Apr 2003	4 Feb 2003
14 Aug 2003	3 Jun 2003
9 Dec 2003	27 Sep 2003
29 Mar 2004	17 Jan 2004
27 Jul 2004	14 May 2004
21 Nov 2004	9 Sep 2004
12 Mar 2005	29 Dec 2004
9 Jul 2005	26 Apr 2005
3 Nov 2005	23 Aug 2005
24 Feb 2006	12 Dec 2005
20 Jun 2006	9 Apr 2006
17 Oct 2006	7 Aug 2006
7 Feb 2007	26 Nov 2006
2 Jun 2007	21 Mar 2007
30 Sep 2007	20 Jul 2007
22 Jan 2008	9 Nov 2007
13 May 2008	4 Mar 2008
10 Sep 2008	1 Jul 2008
4 Jan 2009	22 Oct 2008
26 Apr 2009	14 Feb 2009
25 Aug 2009	13 Jun 2009
18 Dec 2009	6 Oct 2009
9 Apr 2010	27 Jan 2010
6 Aug 2010	26 May 2010
2 Dec 2010	20 Sep 2010

Greatest elongation
of Mercury. (Holley
Bakich)

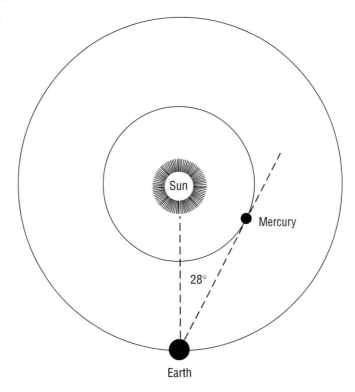

Future transits of Mercury

Date	Time (ET)[a]	Duration
15 Nov 1999	21^{h}43^m	00^{h}52^m
7 May 2003	07^{h}53^m	05^{h}18^m
8 Nov 2006	21^{h}42^m	04^{h}58^m
9 May 2016	15^{h}00^m	07^{h}30^m
11 Nov 2019	15^{h}22^m	05^{h}29^m
13 Nov 2032	08^{h}58^m	04^{h}26^m

Note:
[a] All times are geocentric.

Future close conjunctions of Mercury and the visible planets

	Close conjunctions	
	Date	Separation in declination
with Venus		
	28 Apr 2000	21'
	21 Jun 2003	25'
	14 Jan 2005	21'
	27 Jun 2005	05'
with Mars		
	19 May 2000	67'
	10 Aug 2000	5'
	25 Jul 2002	40'
	10 Jul 2004	10'
	29 Sep 2004	51'
with Jupiter		
	8 May 2000	52'
	20 Jul 2002	75'
	26 Jul 2003	23'
	28 Sep 2004	40'
	6 Oct 2005	88'
with Saturn		
	2 Jul 2002	14'
	1 Jul 2003	93'
	26 Jun 2006	85'

Recent data

Most of the scientific findings about Mercury come from the Mariner 10 spacecraft which was launched on 3 November 1973. It flew past Mercury three times. The first passage took place on 29 March 1974 at a distance of 705 km from the surface. On 21 September 1974 it flew past Mercury for the second time and on 16 March 1975 for the third time. During these visits, over 2700 pictures were taken, covering 45% of Mercury's surface.

As was discussed in a previous section ("The interior composition of Mercury") until Mariner 10 scientists did not suspect that Mercury had a magnetic field. We now know that Mercury's magnetic field is 1% as strong as Earth's. The field is inclined 7° to Mercury's axis of rotation and produces a weak magnetosphere around the planet.

The pictures returned from the Mariner 10 spacecraft showed a world that resembles the Moon. Mercury is covered with craters, contains huge multi-ring basins, and possesses many lava flows. The craters range in size from 100 m (the smallest resolvable feature on Mariner 10 images) to 1300 km. They are in various stages of preservation. Some are young with sharp rims and bright rays extending from them. Others are highly

Mercury from Mariner 10. (NASA)

degraded, with rims that have been smoothed by the bombardment of meteorites.

The largest feature on Mercury is the Caloris Basin. The Caloris Basin is 1300 km in diameter, and was probably caused by a projectile larger than 100 km in size. The impact produced concentric mountain rings 3 km high and sent ejecta 600 to 800 km across the planet. The seismic waves produced from the Caloris impact focused onto the other side of the planet and produced a region of chaotic terrain. After the impact the crater was partially filled with lava flows.

Mariner 10 also showed that Mercury is marked with great curved cliffs or lobate scarps. These were apparently formed billions of years ago as Mercury cooled and shrank a few kilometers in diameter. This shrinking produced a wrinkled crust with scarps kilometers high and hundreds of kilometers long.

The majority of Mercury's surface is covered by plains. Much of the surface is old and heavily cratered, as was expected. However, some of the plains contain far fewer craters per unit area. Scientists have classified these less heavily cratered plains into two types: intercrater plains and smooth plains.

Intercrater plains are less saturated with craters and the craters are less than 15 km in diameter. These plains were probably formed as lava flows buried the older terrain. The smooth plains are younger still with fewer craters. Smooth plains can be found around the Caloris Basin. In some areas it can be

A mosaic of Mariner 10 Mercury images. (NASA)

seen that patches of smooth lava have filled craters some time in Mercury's past.

In 1965, G. H. Pettengill and R. B. Dyce determined Mercury's period of rotation to be 59 ± 5 days based upon radar observations. Later, in 1971, D. Goldstein refined the rotation period to be 58.65 ± 0.25 days using radar observations. After close observation by the Mariner 10 spacecraft, the period was determined to be 58.6461 ± 0.005 days.

During Mercury's distant past, its period of rotation may have been faster. Scientists speculate that its rotation could have been as rapid as 8 hours, but over millions of years it has been slowly despun by solar tides. A model of this process shows that such despinning to the rate at which it rotates now would take 10^9 years and would have raised the interior temperature by 100 K.

Since Mariner 10, some research directed toward Mercury has continued. In 1991, scientists at the California Institute of Technology bounced radio waves off Mercury and found an unusual bright return from the north pole. The apparent brightening at the north pole could be explained by ice on or just under the surface. Researchers wondered if it was possible for Mercury to have ice. Because Mercury's rotation is almost perpendicular to its orbital plane, when Mercury's north pole sees the Sun, it is always just above the horizon. Thus, the insides of craters would never be exposed to the Sun and scientists suspect that they would remain colder than $-161°C$. These sub-freezing temperatures could trap water outgassed from the planet, or ice brought to the planet from cometary impacts. These ice deposits might be covered with a layer of dust but would still show bright radio wave returns.

Historical timeline	c. 385 BC	The Greek astronomer Heraclides became the first person to suggest that Mercury (and Venus) orbit the Sun.
	28 May 1607	Johannes Kepler, using pinhole projection, observed a dark spot against the Sun's disk. He mistakenly labeled this a transit of Mercury and went so far as to inform King Rudolf II.
	7 November 1631	Using predictions made by Johannes Kepler the French astronomer Pierre Gassendi became the first to observe a transit of Mercury.
	3 May 1661	Johannes Hevelius and Christiaan Huygens independently observed a transit of Mercury. Hevelius had predicted that the transit would occur between 1–10 May 1661.
	1682–1684	During this period, British astronomer Edmund Halley made daytime observations of Mercury.
	28 May 1737	Venus occulted Mercury for 2 minutes 7 seconds. Mercury was completely covered from 21:46:44 UT to 21:48:54 UT. The only person on Earth to observe this was the British amateur astronomer John Bevis.
	1878	US astronomer Samuel Pierpont Langley was the first to see Mercury against the solar corona. Langley used the 33-cm refractor at the Allegheny Observatory in Pittsburgh, PA, and noted an angular diameter of 15″ for Mercury just prior to its transit across the solar disk.
	1889	William Harkness measured the mass of Mercury. Giovanni Schiaparelli announced the rotational rate of Mercury as 88 days.
	1962	The first radar contact with Mercury was established by the radio telescope at Arecibo, Puerto Rico.
	1965	G. H. Pettengill and R. B. Dyce, working at the 305-m radio telescope at the Arecibo Ionospheric Observatory in Puerto Rico, determined the rotational period of Mercury to be 59 ± 5 days, based upon radar observations. They beamed a series of 0.005-s and 0.0001-s pulses at Mercury at a frequency of 430 megahertz.

1971	D. Goldstein refined the rotational rate of Mercury to 58.65 ± 0.25 days.
3 November 1973	The US spacecraft Mariner 10 was launched toward Mercury via Venus.
29 March 1974	Mariner 10 encountered Mercury for the first time.
21 September 1974	Mariner 10's second Mercury encounter.
16 March 1975	The third and final encounter of Mercury by Mariner 10.

Venus

Physical data

Size   12104 km

Mass 4.869×10^{24} kg

Escape velocity   10.36 km/s (37296 km/hr)

Temperature range   At the tops of the clouds of Venus, the temperature is approximately $-45\,°C$. At the surface, the temperature is approximately $500\,°C$. There is no diurnal temperature fluctuation as there is on Earth. The official "average" temperature of Venus (as given by NASA) is 737 K (464 °C).

Oblateness*   0

Surface gravity   8.87 m/s^2

Volume 9.284×10^{11} km^3
85.4% that of Earth

Magnetic field strength and orientation   Venus has little or no magnetic field of its own. This is almost certainly due to the slowness of its rotation. A sizable magnetic field is induced in the upper atmosphere (ionosphere) by the solar wind.

Albedo (visual geometric albedo)   0.65

Density (water = 1)   5.25 kg/m^3

Solar irradiance   2660 watts/m^2

Atmospheric pressure   9321900 pascals

Composition of atmosphere

Carbon dioxide (CO_2)	>96.4%
Nitrogen (N_2)	>3.4%
Sulfur dioxide (SO_2)	0.015%
Argon (Ar)	0.007%
Water vapor (H_2O)	0.002%
Carbon monoxide (CO)	0.0017%
Helium (He)	0.0012%
Neon (Ne)	0.0007%
Carbonyl sulfide (COS)	0.00044%
Hydrogen chloride (HCl)	0.00004%
Hydrogen fluoride (HF)	0.000001%

Maximum wind speeds   350 km/hr[†]

---

* There is no appreciable difference between the equatorial diameter of Venus and its polar diameter. This is almost certainly due to the slowness of the planet's rotation. Note: the term "oblateness" is sometimes referred to as a planet's "ellipticity."

[†] This wind speed was measured at the tops of the Venusian clouds. Near the surface, wind speeds are very low, in the range of 0.3 to 1.0 m/s.

Venus in ultraviolet light showing the Y-shaped feature. (NASA)

Outstanding cloud and surface features

Cloud features

Viewed in visible light, there are no permanent features discernible in the clouds of Venus. The atmosphere is in a continuous state of mixing, and any patterns observed quickly dissipate. Through ultraviolet filters, however, a number of markings can be seen. These are visible as immense C- or Y-shaped features which are centered on and symmetrical with the planet's equator. Although individually these features are short-lived, they tend to reform often enough to perhaps be considered a feature in the clouds of Venus. The mechanism by which ultraviolet light is absorbed is not well understood, but it has allowed scientists to gain knowledge regarding the upper atmospheric flow of Venus.

Surface features

Venus, with three times the land area of Earth, has a vast number of large features. It would be impractical to name them all; however, the following list will provide a sample from the dominant types. The major landforms on Venus are in general much larger than their counterparts here on Earth. Diameters for features such as coronae are the "flow apron" diameters (the diameter of the feature subsequent to volcanic activity) rather than the "edifice" diameter (the diameter of the feature prior to any volcanic activity), which would be somewhat less.

Feature	Diameter (km)
Akkruva Colles	1059
Artemis Chasma	3087
Artemis Corona	2600
Aspasia Patera	200
Atanua Mons	1000
Badb Linea	1750
Baker Crater	109
Baltis Vallis	6000
Boadicea Patera	220
Chernava Colles	1000
Citlalpul Valles	2350
Cleopatra Crater	105
Cochran Crater	100
Dali Chasma	2077
Eistla Regio	8015
Fortuna Tessera	2801
Ganiki Planitia	5158
Guinevere Planitia	7519
Haasttse-baad Tessera	2600
Hecate Chasma	3145
Heng-o Corona	1060
Isabella Crater	175
Kaiwan Fluctus	1200
Kalaipahoa Linea	2400
Klenova Crater	141
Maat Mons	280
Maxwell Montes	797
Mead Crater	270
Meitner Crater	149
Morrigan Linea	3200
Mylitta Fluctus	1250
Niobe Planitia	5008
Ovda Regio	5280
Rosa Bonheur Crater	104
Sacajawea Patera	233
Stanton Crater	107
Sudenitsa Tessera	4200
Tellus Tessera	2329
Unelanuhi Dorsa	2600
Vaidilute Rupes	2000
Var Mons	1000
Vedma Dorsa	3345

Orbital data

Period of rotation*   243.0187 days – retrograde
5832.6 hours
$243^{d}0^{h}36.5^{m}$

Period of revolution (sidereal orbital period)   0.61521 years
224.701 days
$0^{y}224^{d}16.8^{h}$

Synodic period $583^{d}22^{h}05^{m}$

Equatorial velocity of rotation   6.520 km/hr

Velocity of revolution   35.02 km/s (126 072 km/hr)

Distance from Sun	Average	0.7233 AU
		108 200 000 km
	Maximum	0.7282 AU
		108 934 900 km
	Minimum	0.7184 AU
		107 463 300 km

Distance from Earth	Maximum	1.75 AU
		261 039 880 km
	Minimum	0.26 AU
		38 150 900 km

Apparent size of Sun (average)   0.74°

Apparent brightness of Sun[†] $m_{vis} = -27.35$

Orbital eccentricity   0.0068

Orbital inclination   3.394°

Inclination of equator to orbit   177.36°

Observational data

Maximum angular distance from Sun   47°19′

Greatest brilliancy[††]   −4.4

| **Angular size**[‡] | maximum | 64″ |
| | minimum | 10″ |

* Due to the rotation of Venus being retrograde and the period of rotation being slightly longer than the period of rotation, the length of a day on Venus is 115 days 16 hours 5 minutes.
† From Venus, the Sun would look slightly larger than from Earth (a little more than twice the surface area), but not appreciably brighter (less than a magnitude). This makes perfect sense, since the Sun's apparent brightness from Earth is already a staggering −26.7. However, the Sun can never even be glimpsed from the cloud-shrouded surface of Venus.
†† Greatest brilliancy for Venus occurs when the planet lies at an elongation of 39°, approximately 36 days before and after inferior conjunction. At this time, the angular size of the planet is approximately 50 seconds of arc.
‡ This measurement is the apparent angular diameter of Venus, measured in seconds of arc, as seen from Earth.

The orbit of Venus.
(Holley Bakich)

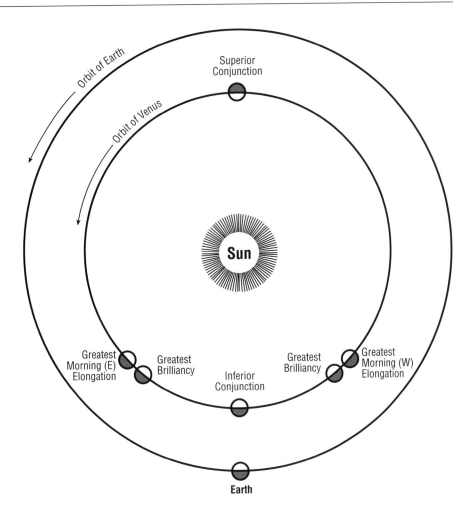

Early ideas

One of the oldest surviving astronomical documents comes to us from the first Babylonian dynasty, being at least as old as 1600 BC. It is from the library of Ashurbanipal and is a series of observational time intervals relating to Venus, which the early Babylonians called Nindaranna. This tablet is a 21-year record which refers to the appearances and disappearances of Venus in the morning and evening skies – giving the correct time intervals – and it goes on to discuss the omens such sightings mean.

The Assyrians called Venus Dil-bat, or Dil-i-pat. In Egypt it was the special star of Ishtar, the mother of the gods. The Chinese knew it as Jin xing. Early in their history, the Greeks thought the evening and morning appearances of Venus were two different objects. They called the planet Hesperus when it appeared in the western evening sky and Phosphorus when it appeared in the eastern morning sky. They soon came to realize that both objects were the same planet.

In the fourth century BC, the Turkish-born Greek astronomer Heraclides Ponticus, a contemporary of Aristotle, put forth an idea related to the placement of Venus (along with Mercury) in the solar system. Up to that time, most thought was geared to an immobile Earth with the Sun, Moon and planets revolving around it. There had always been some thought that perhaps Mercury and Venus did not follow this norm. Heraclides became the first to suggest that Venus (along with Mercury) traveled in circles around the Sun, and not around the Earth.

The name we now employ for the planet Venus comes down to us from the time of the Roman Empire, where Venus was the equivalent of the Greek Aphrodite, the goddess of beauty, love, laughter and marriage. At about the same time, the pseudoscience of astrology was being formalized. According to astrology, Venus is a feminine planet, like the Moon. "She" rules the love nature, children, pleasure, luck, wealth and art. The positive side of Venus (that is, if Venus is well placed in a horoscope) is said to influence our ability to attract, to love, and to express our artistic talents. The negative side of Venus (that is, if Venus is badly placed in a horoscope) affects the emotional nature, producing weaklings prone to erotic, frivolous pastimes and rather pretentious displays of bad taste.

Venus was the most important celestial body observed by the ancient Mayas, who called it Chak ek, the Great Star, a representation of the man-god Quetzalcoatl/Kukulkan. Surprisingly, there are no indications that the Mayas worshipped any of the other planets. Unlike the Greeks, Mayan observers recognized the morning and evening appearances of Venus as the same object. Religious structures were constructed in honor of this object and many were aligned to the most northerly and southerly rising and setting positions of Venus. The Mayas watched Venus with great interest and care. Indeed, their measurement of its synodic year (the time from one appearance of Venus as the evening star to the next) was in error by less than a tenth of a day!

In the fourteenth century, the *Ymago mundi* of Pierre d'Ailly listed the properties of Venus that supposedly related to the influences it had here on Earth: Venus is hot and wet, most splendid amongst the stars, and always companion to the Sun, called Lucifer when it precedes the Sun and Vesper when it follows the Sun.

In the fifteenth century, many accurate measurements of Venus were made by the Persian astronomer Ulugh Beg, grandson of the Mongol conqueror Tamarlane. He built an impressive observatory in the city of Samarkand and used a sextant, 18.3 m in radius, to make detailed observations of the planets and the stars.

From 1582–88, the Danish astronomer Tycho Brahe made a large number of daytime measurements of Venus. He then compared the position of Venus to the Sun, and then that night (or in the morning) he compared Venus's position to specific stars to obtain the most accurate positions up to that time. Earlier observers had used the Moon as a comparison, as it is easily visible in the daytime. This led to greater inaccuracies as the Moon has a rather fast motion through the stars (approximately 13° per day) as seen from the Earth.

In December 1610, the Italian astronomer Galileo Galilei became the first to observe the phases of Venus. This had been predicted by a few individuals who supported the Copernican view of the solar system. (In 1543, Nicolas Copernicus had published *De Revolutionibus Orbium Coelestium*, in which he

Venus. Bianchini, Francesco, *Hesperi et Phosphori nova phaenomena sive observationes circa planetam Veneris*, Rome, 1728. (Linda Hall Library)

placed the Sun at the center of the solar system.) Galileo stated that Venus imitated the Moon in appearance. In addition, Galileo also saw that when Venus was nearly full it was small. That view contrasted with the large apparent size of the planet when he observed Venus as a thin crescent. This difference in apparent size, along with the phases Venus went through, was the strongest possible observational evidence for the validity of the Copernican theory.

Venus, with a spot suggestive of a satellite. Fontana, Francesco, *Novae coelestium terrestriumq[ue] rerum observationes*, Naples, 1646. (Linda Hall Library)

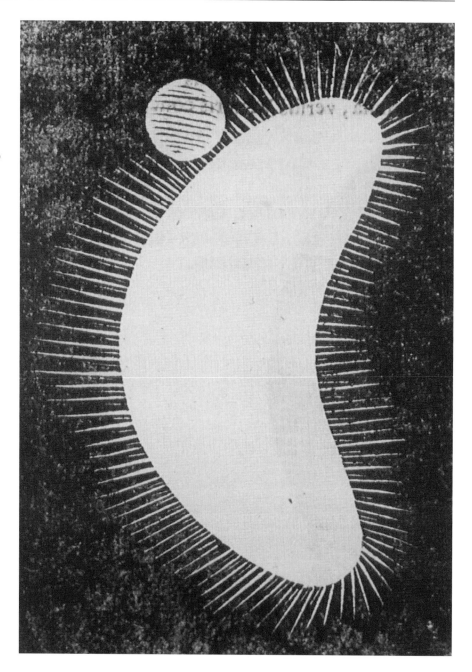

In 1666, the French astronomer Jean-Dominique Cassini made the first measurements of the rotation rate of Venus. He obtained a value of 23 hours and 21 minutes. Cassini also reported sighting a moon in orbit around Venus. This "discovery" had first been reported by the Italian astronomer Francesco Fontana in 1645.

The German amateur astronomer Johann H. Schroeter claimed to have seen

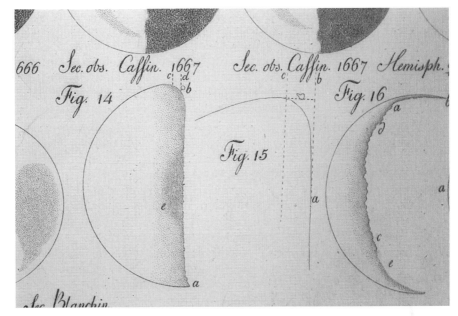

Schroeter's famous illustration of Venus with blunt cusps. Schroeter, Johann, *Aphrodito-graphische Fragmente, zur genauern Kenntniss des Planeten Venus*, Helmstedt, 1796. (Linda Hall Library)

one of the horns of a crescent Venus blunted. He attributed this to the presence of a high mountain. Sir William Herschel was an opponent of such unverified claims and discounted Schroeter's observations, although he did try to verify them. The Italian astronomer Francesco Bianchini made a great number of drawings of Venus, compiling them into a book published the year before his death.

It was during the nineteenth century that the study of Venus began in earnest. Many observers gave a value of nearly 24 hours for the rotation of Venus. The Italian astronomer Giovanni Schiaparelli, however, disagreed with these results. His investigations led him to believe that Venus always kept the same side toward the Sun, the same conclusion he had reached about Mercury.

The nineteenth century also saw a heightened interest in Venus due to the occurrences of two transits of the planet across the bright disk of the Sun. The first of these, in 1874, saw a worldwide effort to try to accurately determine the value of the solar parallax, that number which would provide the exact Earth–Sun distance and thereby furnish astronomers with the scale of the solar system. Russia led the astronomical parade by sending out twenty-six expeditions of astronomers. England sent out twelve, America eight, France and Germany six each, Italy sent three and the Netherlands contributed one. The results of all these endeavors were, to be kind, disappointing. In the words of the leader of one of these expeditions, ". . . observers side by side, with adequate optical means, differed as much as twenty or thirty seconds in the times they recorded for phenomena which they have described in almost identical language." The results were somewhat better for the transit of 1882.

In 1911, the US astronomer Vesto M. Slipher determined by spectral analysis that the rotation rate of Venus was much greater than one day. The rotation rate of Venus had been a major planetary question up until this study was completed. The problem arises from the correspondence between the length of Venus's synodic period (the time interval between successive inferior conjunctions of Venus) and the rotational period of the planet. It works out as 5.001 Venusian days between one inferior conjunction and the next. Since Venus is best observed near inferior conjunction, it means that the same region of Venus is being seen by observers on Earth for a large number of such events.

Important concepts

The greenhouse effect

It is very hot on Venus. In fact, the surface is so hot that for nearly a decade planetary scientists were at a loss to understand what was causing the apparent super-heating. Certainly, since Venus is closer to the Sun than the Earth is, there should be a temperature difference. But when the surface temperature was measured at $464\,°C$, there was a great deal of consternation as to the cause. It was known that there was no large internal source of heating, as Venus is not radiating more energy than it receives from the Sun.

In 1960, the American astronomer Carl Sagan, then a graduate student at the University of Chicago, was the first to calculate the effect of the Venusian atmosphere with respect to "greenhouse" heating, what is now termed the "greenhouse effect." It had been known for some time that climate on the Earth is affected by the greenhouse effect. On Earth, water vapor (H_2O) is the major contributor to this process. The atmosphere of Venus, however, is very dry – water vapor accounts for only 2 parts in 100000. The major contributor to the Venusian greenhouse effect is carbon dioxide (CO_2).

A number of planetary scientists have theorized that Venus once had watery oceans, just like Earth. The planet may have been in equilibrium at that time. Then, something happened. Possibly, an increased output of solar radiation or perhaps something triggered an increase in the CO_2 content of the Venusian atmosphere. The imbalance would create a runaway greenhouse effect which would evaporate water from the oceans and release CO_2 from the surface rocks. As more water evaporated and more CO_2 was released, the temperature would be driven ever higher.

Eventually, ultraviolet radiation from the Sun would break apart the water vapor molecules in Venus's atmosphere. Over time, much of the hydrogen released in this manner would escape from the gravitational pull of Venus. The oxygen, which is quite a bit heavier, would combine with chemicals in the soil and surface rock.

A method to test this idea was first proposed by Thomas Donahue of the University of Michigan. He stated that if hydrogen had been escaping from Venus for a large span of time, the small amount of hydrogen remaining should show a higher percentage of deuterium, an isotope of hydrogen. (Normal hydrogen has a nucleus composed of a proton around which an electron orbits. Deuterium adds a neutron to the nucleus, thereby doubling the atomic weight. Because it is heavier, deuterium escapes into space much

The greenhouse effect. Some sunlight is reflected, but a portion passes through the clouds and is absorbed by the surface and given off as infrared radiation (heat), which does not easily penetrate the cloud layer. (Holley Bakich)

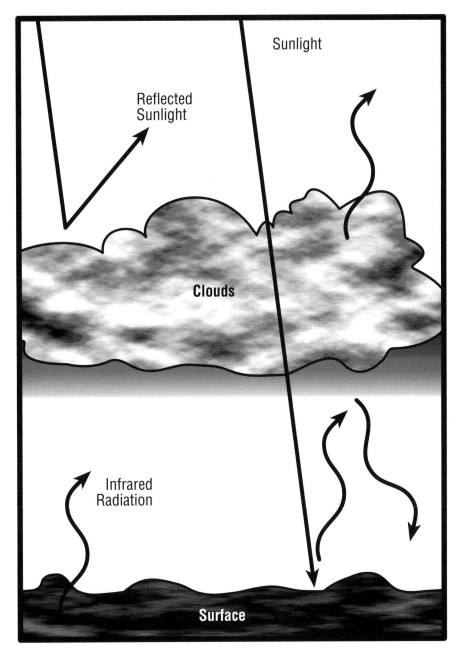

more slowly than normal hydrogen.) All that was needed to test this theory was to measure the ratio of deuterium to hydrogen in the atmosphere of Venus.

On 9 December 1978, one of the probes from the Pioneer Venus spacecraft did just that. Carrying a device used to measure the composition of a planet's atmosphere, called a mass spectrometer, the probe measured for the first time the elusive deuterium/hydrogen ratio. In the oceans of Earth, there is approximately one deuterium atom for every 6000 normal hydrogen atoms. In the atmosphere of Venus, this ratio is one in fifty!

This evidence suggests to some that Venus did indeed possess watery oceans in the distant past. Alone, it is not ultimate proof, but it is one way to explain the massive abundance of carbon dioxide in the atmosphere of Venus, as well as the elevated surface temperature.

The actual mechanism of the greenhouse effect on Venus involves infrared radiation being trapped by the atmosphere. Short wavelength radiation from the Sun reaching the surface of Venus is absorbed and then re-radiated in the form of heat (or infrared) radiation. The carbon dioxide in Venus's atmosphere is essentially opaque to infrared radiation, so it absorbs the heat and then re-emits it. Eventually, a state of equilibrium is reached between the amount of radiation absorbed by the surface and the amount of radiation emitted into space. Because of the massive greenhouse effect, however, the temperature at which this equilibrium is reached on Venus is very high. This is the reason for the tremendous surface temperature on Venus.

Transits of Venus

A transit of Venus occurs when the planet crosses in front of the bright disk of the Sun, as seen from the Earth.

The first transit of Venus to be recorded was the one which occurred in the year 1639. Only two observers saw it: the English astronomer Jeremiah Horrocks and his friend, William Crabtree. The transit which had occurred eight years earlier – in 1631 – had gone unobserved, as far as is known.

A transit of Venus cannot be described as a very striking or beautiful spectacle. It can barely be seen with the unaided eye – and then only with a filter – therefore, it is not as fine a sight as a great comet or a meteor shower. Why is it, then, that the transit of Venus was regarded as having such great scientific importance? It is because the transit enabled us to solve one of the greatest problems which has ever engaged the mind of man. It is by the transit of Venus that astronomers attempted to determine the scale on which our solar system was constructed.

It was comparatively easy to learn the shape of the solar system, to measure the relative distances of the planets from the Sun, and even the relative sizes of the planets themselves. From this information, astronomers constructed a map of the solar system. This included the orbits of the planets, their satellites, asteroids and comets. The easy part was to correctly lay out the *relative* sizes of all these objects. One only had to perform the most basic observations for this. But it was not at all easy to accurately assign the correct scale of this map of the solar system.

In 1691, the English astronomer Edmund Halley put forth a method by which this scale could be measured. Halley explained his method of finding the distance to the Sun by using the transit of Venus which would occur in 1761 (or the following transit, set to take place in 1769). The man most responsible

A transit of Venus showing the geometry of all contacts. (Holley Bakich)

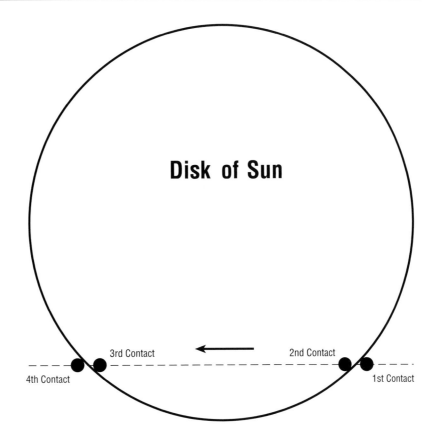

Disk of Sun

3rd Contact

2nd Contact

4th Contact

1st Contact

for organizing the worldwide effort to observe these transits was the French astronomer Joseph-Nicolas Delisle. The results from the first of this pair of transits were not very successful, in spite of the arduous labors of those who undertook the observations.

The transit of 1769, however, will be forever remembered, not only on account of the determination of the Sun's distance, but as giving rise to the first of the celebrated voyages of Captain Cook. These observations were regarded as so important for the sake of all mankind that the French government instructed its men-of-war not to attack the ships of Captain Cook. It was to see the transit of Venus that Captain Cook was commissioned to sail to Otaheite (Tahiti), in what is now French Polynesia. There, on 3 June 1769, a splendid day in that most excellent climate, the transit of Venus was carefully observed and measured by different observers. As Captain Cook's party was making its observations of the transit, other measurements were being obtained in Europe and North America, and from the combination of these the first (reasonably) accurate knowledge of the Sun's distance was determined.

Amazingly, the refinement of these observations did not take place for some time. It was not until 1824 that the German astronomer Johann Franz Encke computed the distance of the Sun. He gave 95 000 000 miles as the definite (albeit incorrect) result. Still, for a number of years, that value was

A nineteenth-century projection of the 2004 transit of Venus. Proctor, Richard Anthony, *Old and New Astronomy*, London, 1892. (Michael E. Bakich collection)

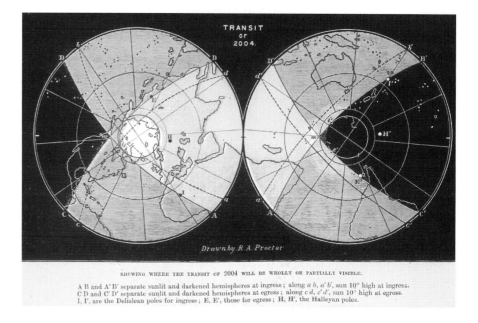

SHOWING WHERE THE TRANSIT OF 2004 WILL BE WHOLLY OR PARTIALLY VISIBLE.

A B and A′ B′ separate sunlit and darkened hemispheres at ingress; along *a b*, *a′ b′*, sun 10° high at ingress. C D and C′ D′ separate sunlit and darkened hemispheres at egress; along *c d*, *c′ d′*, sun 10° high at egress. I, I′, are the Delislean poles for ingress; E, E′, those for egress; H, H′, the Halleyan poles.

adopted. It would take another transit of Venus before the Sun's distance was revised.

The two transits of Venus which occurred in the nineteenth century attained an importance unsurpassed by any observed occurrence in the solar system, and, in fact, received a degree of attention never before accorded to any astronomical phenomenon. Observers all over the world had an army of telescopes aimed at the Sun for these two events. Unfortunately, the results obtained were less than satisfying.

The main problem that arose was the reconciliation of timings of the entrance of Venus onto the disk of the Sun (and, at the end of the transit, its exit from the disk) from one observer to the next. It might be assumed that because Venus is a black circle and the solar disk is bright, the moment when the entire sphere of Venus crosses onto the Sun's face would be easy to determine. In real circumstances, however, the disk of Venus seems almost to attach itself to the limb of the Sun in a similar way to a water droplet emerging from a tap. Then, all at once, the contact is broken and Venus stands some distance in from the solar limb. Nineteenth-century observers christened this phenomenon the "Black Drop" or the "Black Ligament." Due to this situation, the observations of skilled observers with decades of experience showed significant differences in timing this important event.

Still, averaging out these differences allowed nineteenth-century astronomers to refine Encke's calculation of the Sun–Earth distance. Following the two transits of Venus, a figure of 149 182 110 km was published and generally accepted. As can be seen, this number is much closer to the presently accepted value (149 597 892 km) and its calculation was a significant step forward in the refinement of the scale of our solar system.

Transits of Venus have also allowed astronomers to learn two important

The Crescent Moon
and Venus.
(Photograph by the
author)

facts about our "twin" world. First, they learned with a great deal of assurance that Venus had no moon. It had been conjectured by some that if Venus were attended by a small body in close proximity, the brilliancy of the planet would overwhelm the light of any satellite, and thus a moon would remain undiscovered. It was, therefore, a matter of some importance to carefully examine the vicinity of the planet during a transit. If a satellite of any appreciable dimensions had existed, it would have been detected against the brilliant background of the Sun.

Another fact first gleaned about Venus as a result of a transit was the existence of an atmosphere. This was first discovered by M. V. Lomonosov during the transit of 1761. It was thought that if Venus had no atmosphere it would be totally invisible just before beginning its crossing of the solar disk, and would relapse into total invisibility immediately after the transit. The observations proved otherwise, however. As Venus gradually moved off the Sun, the circular edge of the planet extending out into the darkness was seen to be bounded by an arc of light. Some observers, under extremely favorable conditions, have been able to follow the planet until it passed entirely away from the brilliant solar background. At this point the globe of Venus, though itself invisible, was distinctly marked by the circle of light surrounding it. The only explanation possible was that Venus was surrounded by an atmosphere.

How the Sun's distance is measured by a transit of Venus

To understand the importance of a transit of Venus to nineteenth-century astronomers, we must first understand the concept of parallax. In fact, a

The twenty-first-century transits of Venus. For times, see page 125. (Holley Bakich)

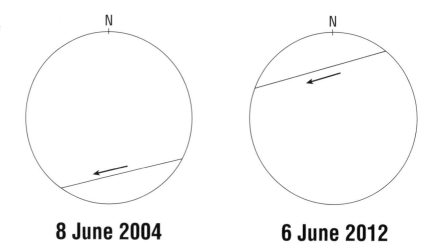

8 June 2004　　　　　**6 June 2012**

number of astronomy books – both old and new – define the measurement of the Earth–Sun distance as the calculation of the "solar parallax." Parallax is defined as the apparent displacement of an object against a background when the object is viewed from different locations. If you hold your thumb at arm's length and alternately close first one eye and then the other, the position of your thumb seems to change against the background objects. This apparent shift is called parallax.

Parallax is most often applied to the stars to determine their distance. Stellar parallax is the apparent angular displacement of a star that results from the revolution of the Earth about the Sun. To measure the parallax of a nearby star, its position is noted. Six months later (when the Earth is on the opposite side of the Sun), the star's position is again measured. The angle of the star image's apparent shift against the background stars is measured. This is the parallax of the star. Since the angles measured are extremely small, parallax is useful for obtaining stellar distances out to approximately 100 parsecs (about 326 light years). Numerically, parallax is an angle measured in seconds of arc. One second of arc equals 1/3600 of a degree.

Relating a transit of Venus to the first example above, we substitute two observers on Earth for our two eyes, Venus takes the place of our thumb, and the brilliant disk of the Sun becomes the background. When Venus is observed by two observers separated by as wide a distance as is possible (or practical), the amount of displacement may be measured. From that displacement (parallax), we may calculate the distance.

For this procedure to be of value, an accurate measure of timings throughout the transit must be made by all observers. The most important measurements are those of the four "contact" points. First Contact occurs when the edge of the planetary disk of Venus begins its journey across the solar disk. Second Contact is the moment when the entire disk of Venus is seen against the solar background. This occurs when the trailing edge of the planetary disk touches the circumference of the Sun's disk. Similarly, Third Contact and Fourth Contact occur when the disk of Venus touches, and then exits, the brilliant disk of the Sun. As a matter of terminology, First and Fourth Contacts are often referred to as "external" contacts, while Second and Third Contacts are "internal."

Why was the rotational rate of Venus so difficult to determine?

As has been mentioned earlier, the rate at which Venus rotates on its axis is now known with great precision. It was not until 1961, however, that this value was accurately measured. The reason for this is a peculiar mathematical correspondence between the length of the Venusian day (approximately 116 Earth days) and the length of time between successive inferior conjunctions of Venus (approximately 584 Earth days). This works out to 5.001 Venusian days between one inferior conjunction and the next.

In the early days of radio telescopic observations of Venus, this meant that a feature which was observed at the time of one inferior conjunction would be in almost exactly the same position at the time of the next inferior conjunction! Such a multiple periodic resonance was unheard of and very confusing to astronomers. It would be easy to imagine an astronomer, confronted with a radar echo which shows the same terrain as was observed 584 days before, arriving at a variety of conclusions.

It is also easy to conjecture why early telescopic observers suggested such wildly divergent values for the rotational period of Venus. They were attempting to ascertain the rotation of the planet by observing the tops of the clouds. Whatever features or markings they may have imagined they saw were, in actuality, only temporary, evanescent markings in the clouds.

As yet there has been no satisfactory explanation given as to the correspondence between the Venusian day and the time between successive inferior conjunctions of Venus. Tidal forces exerted by the Earth were an early attempt to account for this effect. Such tidal forces are the reason that we on the Earth see only one side of the Moon, while the other side continually faces space. However, calculations have shown that the Earth's tidal forces are much too weak at the distance of Venus to have such an effect.

Interesting facts

The orbit of Venus is the most circular of any planet in the solar system. The eccentricity of its orbit is only 0.0068.

Venus was first observed at radio wavelengths in 1958, although there was an unconfirmed radio observation by J. D. Kraus in 1956.

Altitudes on Venus are given as height above the average global elevation (since there is no "sea level" reference, as on Earth).

The highest peak on Venus resides in the mountain range known as Maxwell Montes, and has an elevation of 11 580 m.

121

Venus. Zahn,
Johann, *Specula
physico-
mathematico-
historica
notabilium ac
mirabilium
sciendorum*,
Nuremberg, 1696.
(Linda Hall Library)

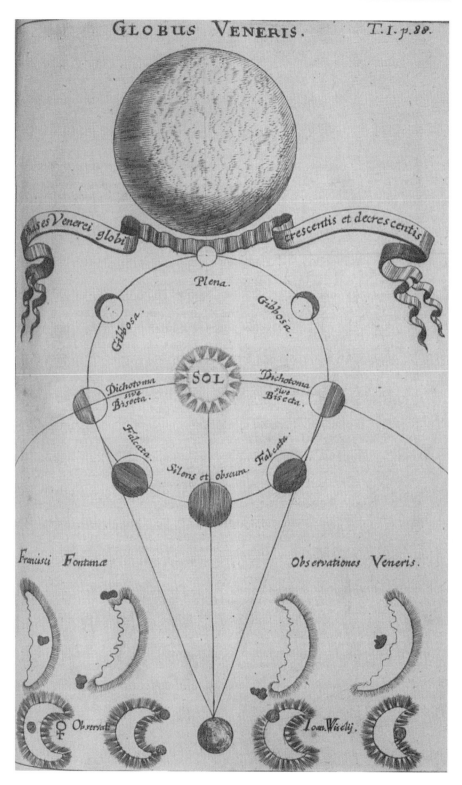

Time-lapse of Venus setting over the Tucson Mountains.
(Photograph by the author)

The first observation of a transit of Venus was by Jeremiah Horrocks and William Crabtree on 4 December 1639 (this was the date on the Gregorian calendar. The date on the Julian calendar, which we now use, is 24 November 1639).

For an observer on Venus, the Earth at opposition would have an apparent visual magnitude of −6.7. The Moon would have an apparent visual magnitude of −2.5. The maximum separation of the Earth–Moon pair would be 31.9 arc-minutes.

Due to the thickness of the atmosphere of Venus, causing meteors to decelerate as they fall toward the surface, no impact crater smaller than about 3.2 km across can form.

The rotation rate of Venus was discovered in 1961, at the 64-m Goldstone Radio Telescope in the Mojave Desert of California.

R. L. Carpenter, D. Goldstein, F. Drake and O. N. Rzhiga reported a retrograde rotation for Venus in 1962.

The retrograde rotation of Venus was confirmed by workers at the Arecibo Ionospheric Observatory, Arecibo, Puerto Rico, in 1964.

Maat Mons is the highest volcano on Venus.

Venus was first shown to be an emitter of microwave radiation in 1956.

The temperature at the cloud tops of Venus's atmosphere is $-45\,°C$.

The first individual given credit for recognizing that the morning and evening appearances of Venus were the same object was the Greek philosopher and mathematician Pythagoras.

Discovery of an atmosphere surrounding Venus was made by M. V. Lomonosov and occurred during the planet's transit in 1761.

The high peaks of Venus, that is, those with elevations over 3960 m, are surprisingly reflective, much more so than the surface at lower elevations. Although ideas abound, the reason for this is as yet unclear.

Maxwell Montes is the only feature on Venus not named after a female.

In 1737 Venus passed directly in front of Mercury, as seen from Earth. This was observed by the English astronomer John Bevis at the Royal Observatory, Greenwich, England.

Venus has more dry land than any other planet in our solar system. It has three times the dry land area of Earth.

Nearly 90% of the surface of Venus is covered by volcanic landforms.

Tesserae (very rugged cracked and wrinkled land areas) occupy approximately 10% of the landscape of Venus.

Greatest brilliancy for Venus (as seen from Earth) occurs when the planet lies at an elongation of 39°, approximately 36 days before and after inferior conjunction.

Venus has a total of 1761 named features, of which 870 are craters.

The largest crater on Venus is named Mead, with a diameter of 270 km.

If you could observe the Sun from the surface of Venus, it would be 1.8 times as bright as from the Earth.

The atmospheric pressure on Venus (9 321 900 pascals) is equivalent to being 914 m under the surface of the Earth's oceans.

Observing data   *Future dates of conjunction*

Dates of conjunction	
Inferior	Superior
20 Aug 1999	11 Jun 2000
30 Mar 2001	14 Jan 2002
31 Oct 2002	18 Aug 2003
8 Jun 2004 (transit)	31 Mar 2005
14 Jan 2006	28 Oct 2006
18 Aug 2007	9 Jun 2008
28 Mar 2009	11 Jan 2010
29 Oct 2010	

Future greatest elongation dates

East of Sun (evening sky)	West of Sun (morning sky)
11 Jun 1999	30 Oct 1999
17 Jan 2001	8 Jun 2001
22 Aug 2002	11 Jan 2003
29 Mar 2004	17 Aug 2004
3 Nov 2005	25 Mar 2006
9 Apr 2007	29 Oct 2007
14 Jan 2009	7 Jun 2009
20 Aug 2010	

Future transits of Venus

Date	Time (ET)[a]	Duration
8 Jun 2004	08^h24^m	06^h12^m
6 Jun 2012	01^h36^m	06^h20^m
11 Dec 2117	02^h51^m	05^h40^m
8 Dec 2125	16^h01^m	05^h32^m
11 Jun 2247	11^h43^m	05^h44^m
9 Jun 2255	04^h50^m	07^h00^m

Note:
[a] All times are geocentric. This is the moment of mid-transit.

Future close conjunctions of Venus and the visible planets

	Close conjunctions	
	Date	Separation in declination
with Mercury		
	28 Apr 2000	21'
	21 Jun 2003	25'
	14 Jan 2005	21'
	27 Jun 2005	05'
with Mars		
	21 Jun 2000	18'
	10 May 2002	18'
	5 Dec 2004	75'
with Jupiter		
	17 May 2000	01'
	5 Aug 2001	72'
	3 Jun 2002	99'
	21 Aug 2003	34'
	4 Nov 2004	36'
	2 Sep 2005	82'
with Saturn		
	18 May 2000	74'
	15 Jul 2001	44'
	8 Jul 2003	49'
	25 Jun 2005	78'

Recent data

All of the recent data about Venus was obtained by the terrifically successful NASA Magellan mission. The Magellan spacecraft, named after the sixteenth-century Portuguese-born explorer whose expedition first circumnavigated the Earth, was deployed from the space shuttle Atlantis on May 4 1989, and arrived at Venus on August 10 1990. Magellan's solid rocket motor placed it into a near-polar elliptical orbit around the planet with a periapsis altitude of 294 km at 9.5° N. The spacecraft is topped by a 3.7 m-diameter dish-shaped antenna that was a spare part left over from the Voyager program.

Designed as a follow-up to the mapping portion of the Pioneer Venus mission, Magellan's purposes were to: (1) obtain near-global radar images of Venus's surface with a resolution equivalent to optical imaging of 1 km per line pair; (2) obtain a near-global topographic map with 50 km spatial and 100 m vertical resolution; (3) obtain near-global gravity field data with 700 km resolution and 2–3 milligals (1 gal = 1 cm/s^2) accuracy; and (4) develop an understanding of the geological structure of the planet, including its density distribution and dynamics.

During the first 8-month mapping cycle around Venus, Magellan collected

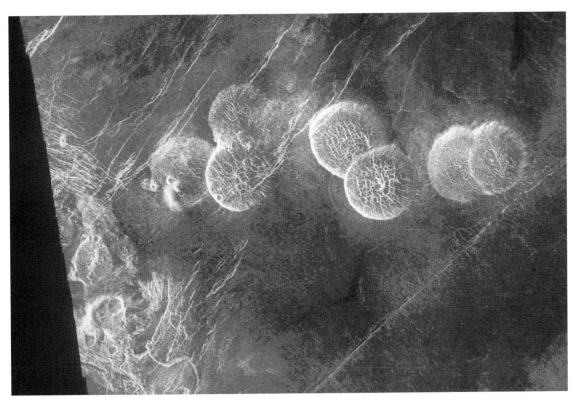

The Magellan spacecraft imaged pancake-like structures on Venus. (NASA)

radar images of 84% of the planet's surface, with resolution 10 times better than that of the earlier Soviet Venera 15 and 16 missions. Altimetry and radiometry measurements were also made, yielding information about the surface topography and electrical characteristics.

During the extended mission, two further mapping cycles from May 15 1991 to September 14 1992 brought mapping coverage to 98% of the planet. A total of 4225 usable imaging orbits was obtained by Magellan. Each orbit typically covered an area 20 km wide by 17000 km long, at a resolution of 120 m/pixel. Image pixels were later rescaled to 75 m/pixel.

Precision radio tracking of the spacecraft measured Venus's gravitational field to show the planet's internal mass distribution and the forces which have created the surface features. An aerobraking maneuver circularized the orbit to improve gravity measurements.

The Magellan mission's scientific objectives were to study landforms and tectonics, impact processes, erosion, deposition, chemical processes, and model the interior of Venus. Magellan showed us an Earth-sized planet with no evidence of Earth-like plate tectonics. At least 85% of the surface is covered with volcanic flows, the remainder by highly deformed mountain belts. Even with the high surface temperature (464 °C) and high atmospheric pressure (9 321 900 pascals), the complete lack of water makes erosion

Sif Mons, a volcano on Venus, is shown in this Magellan image. The area shown is approximately 800 km across. (NASA)

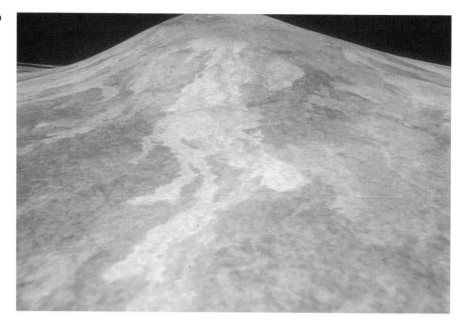

a negligibly slow process, and surface features can persist for hundreds of millions of years. Some surface modification in the form of wind streaks was observed.

Over 80% of Venus lies within 1 km of the mean radius of 6051.84 km. The mean surface age is estimated to be about 500 million years. A major unanswered question concerns whether the entire surface was covered in a series of large events 500 million years ago, or if it has been covered slowly over time. The gravity field of Venus is highly correlated with the surface topography, which indicates the mechanism of topographic support is unlike that of the Earth, and may be controlled by processes deep in the interior. Details of the global tectonics on Venus are still unresolved.

NASA mission scientists have identified seven key scientific results from the Magellan mission:

(1) Study of the Magellan high-resolution global images is providing evidence to understand the role of impacts, volcanism and tectonism in the formation of Venusian surface structures.

(2) The surface of Venus is mostly covered by volcanic materials. Volcanic surface features, such as vast lava plains, fields of small lava domes and large shield volcanoes are common.

(3) There are few impact craters on Venus, suggesting that the surface is, in general, geologically young – less than 800 million years old.

(4) The presence of lava channels over 6000 km long suggests river-like flows of extremely low-viscosity lava that probably erupted at a high rate.

(5) Large pancake-shaped volcanic domes suggest the presence of a type of lava produced by extensive evolution of crustal rocks.

(6) The typical signs of terrestrial plate tectonics – continental drift and basin floor spreading – are not in evidence on Venus. The planet's tectonics is dominated by a system of global rift zones and numerous broad, low domical structures called coronae, produced by the upwelling and subsidence of magma from the mantle.

(7) Although Venus has a dense atmosphere, the surface reveals no evidence of substantial wind erosion, and only evidence of limited wind transport of dust and sand. This contrasts with Mars, where there is a thin atmosphere, but substantial evidence of wind erosion and transport of dust and sand.

The Magellan mission ended when the spacecraft was allowed to plunge into Venus's atmosphere in October 1994. Of course, results from the Magellan mission to Venus are still under investigation. It will take many years for researchers to sort through all the information returned by this extremely valuable spacecraft.

Historical timeline	c. 385 BC	The Greek astronomer Heraclides was the first person to suggest that Venus (and Mercury) orbit the Sun.
	20 May 573	Only 36 days after being occulted by Mars, the star η Cancri was occulted by Venus, as observed by the Chinese.
	30 November 773	The Chinese observed an occultation of β Scorpii by Venus; earlier in the same year – on 4 May – the same star was occulted by Jupiter.
	9 September 885	Seen from China, Venus occulted α Leonis (Regulus).
	December 1610	Galileo observed that Venus progresses through phases similar to those of the Moon.
	4 December 1639	The English astronomer Jeremiah Horrocks became the first person to observe a transit of Venus, along with his friend, William Crabtree.
	1666	The French astronomer Jean-Dominique Cassini made the first measurements of the rate at which Venus spins on its axis. He obtained a value of 23 hours 21 minutes.
	1691	The English astronomer Edmund Halley presented a method by which the distance scale of the solar system may be determined by using a transit of Venus.
	6 June 1761	A transit of Venus was visible in its entirety over Asia and the north polar regions. It was during this transit that the Russian astronomer Mikhail Lomonosov, observing from the University Observatory in St Petersburg, made the discovery that Venus has an atmosphere.

3 June 1769	A transit of Venus was visible in its entirety over the Pacific Ocean, western America and the north polar regions.
3 January 1818	At 21:51 ET, Venus occulted Jupiter. There are no reports of this event being observed. This was the last mutual occultation by planets until 22 November 2065, when, at 12:47 ET, Venus will again pass in front of Jupiter.
1824	The German astronomer Johann Franz Encke, using data collected during the 3 June 1769 transit of Venus, calculated the Earth–Sun distance. His figure of 95 000 000 miles became the standard for the next several decades.
9 December 1874	A transit of Venus was visible from North America, the Atlantic Ocean and western Europe. Much hope was placed upon the observations of this transit, which was the first to be photographed. Unfortunately, the data obtained provided widely varying values for the Earth–Sun distance.
6 December 1882	A transit of Venus was visible from Europe and Asia. This transit allowed the scale of the solar system (based upon the average of the Earth–Sun distance) to be refined to an accuracy of 1 part in 357.
1911	The US astronomer Vesto M. Slipher used spectral analysis to show that the rate at which Venus rotates is much longer than one day. Slipher actually began spectral analysis of Venus many years earlier.
1923	First ultraviolet photographs of the cloud features of Venus.
1932	Carbon dioxide (CO_2) was discovered in the atmosphere of Venus by the astronomers T. Dunham and W. S. Adams.
1956	Venus was shown to be an emitter of microwave radiation.
1958	Venus was observed at radio wavelengths.
1961	Workers at the Goldstone Radio Telescope accurately measured the rate at which Venus spins on its axis.
12 February 1961	Venera 1 space probe launched by USSR. Radio failed when the spacecraft was 22 500 000 km from Venus, but the craft was tracked by radar to its closest approach, 19 May 1961, when it was only 100 000 km from the planet. This marked the first launch of a spacecraft which escaped the gravitational pull of Earth.
1962	The US astronomer Carl Sagan, while still a graduate student at the University of Chicago, became the first to calculate the effect of the Venusian atmosphere on the surface temperature of Venus.

14 December 1962	Mariner 2 space probe launched by US. The probe flew by Venus at a distance of 34 275 km. This was the first successful US planetary flyby.
2 April 1964	Zond 1 space probe launched by USSR. This spacecraft passed within 96 500 km of Venus on 19 July 1964.
12 November 1965	Venera 2 space probe launched by USSR. This craft passed within 24 140 km of Venus on 27 February 1966.
16 November 1965	Venera 3 space probe launched by USSR. This spacecraft crash-landed on the surface of Venus on 27 February 1966. On that date, Venera 3 became the first human artifact to reach the surface of another planet.
12 June 1967	Venera 4 space probe launched by USSR. On 18 October 1967, this spacecraft dropped a 383-kg probe into the atmosphere of Venus. This probe descended by parachute for 96 minutes. The probe carried sensors for temperature, pressure, atmospheric density and gas analysis. It transmitted data until it reached an altitude of 24.1 km.
19 October 1967	Mariner 5 space probe launched by US. This spacecraft flew within 4023 km of Venus. It sent back readings on Venus's surface temperature and magnetic field strength.
5 January 1969	Venera 5 space probe launched by USSR. Landed on Venus 16 May 1969. This spacecraft sent back data for 53 minutes while descending through the atmosphere of Venus.
10 January 1969	Venera 6 space probe launched by USSR. It traveled through the atmosphere of Venus on 17 May 1969, one day after the Venera 5 craft had arrived. This spacecraft sent back data for 51 minutes while descending through the atmosphere of Venus.
17 August 1970	Venera 7 space probe launched by USSR. On 15 December 1970, this spacecraft became the first to soft-land on another planet. It transmitted 25 minutes of data during its descent through the Venusian atmosphere and 23 minutes from the night side of the surface. It measured the temperature to be 464 °C and the atmospheric pressure as 90 times that of Earth.
27 March 1972	Venera 8 space probe launched by USSR. This craft reached the surface of Venus, soft-landing on the daytime side on 22 July 1972. Venera 8 contained a soil analyzer and sent back data for 50 minutes.
5 February 1974	Mariner 10 space probe launched by US. It flew by Venus at a distance of 5633 km and then traveled to Mercury. It thus became the first two-planet space probe.

8 June 1975	Venera 9 space probe launched by USSR. On 22 October 1975, after a successful soft-landing on the surface of Venus, this space probe relayed 115 minutes of television pictures. The images showed a planet with a relatively bright, young surface.
14 June 1975	Venera 10 space probe launched by USSR. This craft soft-landed on Venus three days after Venera 9, on 25 October 1975, albeit 2250 km away. It transmitted data on temperature, rock density and wind velocity from the surface of Venus for 65 minutes.
9 September 1978	Venera 11 space probe launched by USSR. The main spacecraft flew by Venus at a distance of 35400 km. The lander touched down on Christmas Day 1978, 800 km from where Venera 12 had landed four days before. It relayed data on temperature, pressure and atmospheric composition back to Earth for 110 minutes.
14 September 1978	Venera 12 space probe launched by USSR. This craft landed softly on the surface of Venus on 21 December 1978. It transmitted data for 95 minutes.
4 December 1978	Pioneer-Venus 1 (also known as Pioneer 12) space probe launched by US. This spacecraft entered into orbit around Venus and mapped the surface via radar to a resolution of 80 km.
9 December 1978	Pioneer-Venus 2 (also known as Pioneer 13) space probe launched by US. This spacecraft released a series of four probes which entered the atmosphere of Venus.
30 October 1981	Venera 13 space probe launched by USSR. On 1 March 1982, this spacecraft landed in a volcanic area and relayed color television pictures of the surface and the sky.
4 November 1981	Venera 14 space probe launched by USSR. This spacecraft landed on Venus on 5 March 1982 and sent back data to Earth for 57 minutes.
2 June 1983	Venera 15 space probe launched by USSR. This craft entered orbit around Venus on 10 October 1983. Its primary function was to map via radar a significant portion of the northern hemisphere of Venus using a 5.5-m antenna. Over 194 million square kilometers of surface were mapped with a resolution of 1–2 km.
6 June 1983	Venera 16 space probe launched by USSR. This spacecraft went into orbit around Venus on 14 October 1983. In addition to some radar mapping of the surface, Venera 16 (in conjunction with Venera 15) compiled a detailed temperature map of the northern hemisphere of Venus.

15 December 1984	Vega 1 space probe launched by USSR. This was a dual probe, incorporating a lander for Venus and an instrument probe for Halley's Comet. The name is a combination of Venus and Halley (Gallei, in Russian). The Venus portion soft-landed on the planet on 11 June 1985 and transmitted for 21 minutes. In addition, while 53 km above the surface, the lander released a French-made instrument package. This gondola supported by balloons radioed data back to Earth for 46 hours.
21 December 1984	Vega 2 space probe launched by USSR. Essentially the same as the Vega 1 probe, the Venus portion soft-landed on the surface of the planet on 15 June 1985.
4 May 1989	Magellan space probe launched by US. This craft entered into orbit around Venus on 10 August 1990, and within 16 months it had radar-mapped over 90% of the surface of the planet. The resolution of these maps is the best ever obtained, with surface details discernible which are only 0.18 km across.
10 February 1990	US–European space probe Galileo, on its way to Jupiter, flew by Venus for a gravity boost.
8 June 2004	A transit of Venus with a duration of 06^h12^m (measured from First Contact to Fourth Contact) will occur. The mid-point of this event will occur at 08^h24^m Greenwich Mean Time. This transit will be best seen from Europe and Asia.
6 June 2012	A transit of Venus with a duration of 06^h20^m (measured from First Contact to Fourth Contact) will take place. The mid-point of this event will occur at 01^h36^m Greenwich Mean Time. This transit will be best viewed from western North America and the Pacific Ocean.

Earth

Physical data

Size   12 756.28 km*

Mass 5.976×10^{24} kg

Escape velocity   11.18 km/s (40 248 km/hr)

Temperature range

	Minimum	Average	Maximum
	$-69.0\,°C$	$7\,°C$	$58\,°C$

Oblateness   0.003 35

Surface gravity 9.78 m/s^2

Volume 1.087×10^{12} km^3

Magnetic field strength and orientation   The magnetic field of the Earth results from electric currents circulating within the liquid core. Planetary rotation induces a dynamo-like effect in the metallic core, caused by slow movements of the materials there. The strength and alignment of the magnetic field vary due to changes in these motions. The extent of the magnetic field into space also changes, but generally it extends 48 000–64 000 km above the Earth's surface in the direction toward the Sun and 322 000–362 000 km in the direction opposite the Sun. The difference is due to the pressure exerted on the magnetosphere by the solar wind.

Average surface field (tesla)	0.00003
Dipole moment (weber-meters)	6×10^{14}
Tilt from planetary axis	11°
Offset from planetary axis (planet radii)	0.0

Albedo (visual geometric albedo)   0.37

Density (water = 1)   5.515 g/cm^3

Solar irradiance   1380 watts/m^2

Atmospheric pressure   101 325 pascals†

Composition of atmosphere

Molecular nitrogen (N_2)	>78.0%
Molecular oxygen (O_2)	>20.9%
Argon (Ar)	0.93%
Carbon dioxide (CO_2)	0.03%
Neon (Ne)	0.002%

Maximum wind speeds   483 km/hr

---

* Polar diameter 12 712 km.
† This is the standard measure, usually assumed to be the average atmospheric pressure of Earth at sea level.

A hurricane seen
from space.
(NASA)

Outstanding cloud and surface features

Cloud features

The Earth is a moderately cloudy planet, although no permanent cloud features are visible from space. Occasionally, vast storms known as tropical cyclones may be seen. Such storms may spawn hurricanes. (Hurricanes are known as typhoons in the Southern Hemisphere.) Because of the Coriolis Force, the clouds in these storms rotate counterclockwise in the Northern Hemisphere and clockwise in the Southern Hemisphere.

Efforts to predict climatic changes associated with global warming have focused new attention on the warming and cooling properties of clouds. The picture is complex. All clouds block some incoming solar radiation, and absorb some of the heat radiated from the Earth's surface. Many atmospheric scientists now think that the lower-altitude cumulus clouds have a net cooling effect on Earth's surface, reflecting heat back to space. Conversely, the higher, thin cirrus clouds trap heat, radiating it back to the surface of Earth.

Recent studies suggest that the cooling effects of great masses of cumulus storm clouds over the ocean at mid-latitudes outweigh the heating effects of the upper-level cirrus clouds when considered on a global scale, somewhat offsetting global warming. Still, many models of global warming predict a decline in heavy mid-latitude cumulus storm clouds in the future. The amount of high-level cirrus cloud is predicted to rise as the cumulus decreases. If these predictions prove true, such changes will in turn induce accelerated global warming.

Surface features

Land area

Continents	Land area	
	km²	% of surface
Asia	44 500 000	8.73
Africa	30 302 000	5.95
North America	24 241 000	4.76
South America	17 793 000	3.49
Antarctica	14 100 000	2.77
Europe	9 957 000	1.95
Australia/Oceania	8 557 000	1.68
Total	149 450 000	29.33

Highest mountains

Mountain	Height (m)
Everest	8848
K2	8611
Kanchenjunga	8598
Lhotse	8516
Makalu	8481
Cho Oyu	8201
Dhaulagiri	8172
Manaslu	8156
Nanga Parbat	8126
Annapurna	8078

Water area

Oceans	Water area	
	km²	% of surface
Pacific Ocean	179 679 000	35.26
Atlantic Ocean	92 373 000	18.13
Indian Ocean	73 917 000	14.51
Arctic Ocean	14 090 000	2.77
Total	360 059 000	70.67

The five largest Seas

Sea	Area (km²)	% of surface
South China Sea	2974600	0.58
Caribbean Sea	2766000	0.54
Mediterranean Sea	2516000	0.49
Bering Sea	2268000	0.44
Gulf of Mexico	1543000	0.30

Orbital data

Period of rotation   0.99727 days
23.9345 hours
$0^d23^h56.1^m$

Period of revolution (sidereal orbital period)   1.0000 year
365.2422 days
$1^y0^d0^h$

Equatorial velocity of rotation   463.83 m/s (1669.8 km/hr)

Velocity of revolution   29.79 km/s (107244 km/hr)

Distance from Sun   Average   1.0000 AU
149597870 km

Maximum   1.0168 AU
152104980 km

Minimum   0.9832 AU
147085800 km

Apparent size of Sun (average)   0.53°

Apparent brightness of Sun $m_{vis} = -26.7$

Orbital eccentricity   0.0167

Orbital inclination*   0°

Inclination of equator to orbit   23.45°

Early ideas

Any attempt to summarize the many early ideas that humanity has had regarding our planet would take a great deal more space than can be allotted here. I have chosen to list a few creation myths followed by several important ideas or discoveries. This is not to say that the hundreds of millions that I have omitted are less important, only that choices had to be made.

In lower Egypt thousands of years ago, Geb was worshipped as the Earth god. According to legend, Geb married his sister Nut, the sky goddess, without the permission of the powerful Sun god Ra. Ra was so angry at Nut

* This parameter is defined by the Earth's orbit.

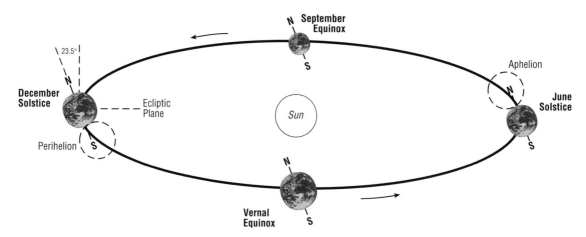

The orbit of the Earth. (Holley Bakich)

and Geb that he forced their father Shu, the god of air, to separate them. That is why the Earth is divided from the sky.

Gaea, or Mother Earth, was the great goddess of the early Greeks. She represented the Earth and was worshipped as the universal mother. In the creation story of the ancient Greeks, the first beings were Chaos and Gaea. From their union sprang the starry heavens, in the form of the sky god Uranus, as well as the mountains, plains, seas and rivers that make up the Earth as we know it today.

To the early Nordic and Germanic tribes, Midgard, the Earth, was created from the flesh of the frost giant Ymir. Odin, the eventual father of the gods, and his brothers Vili and Ve, killed Ymir and formed the ocean with his blood and sweat. His bones and teeth became the mountains and hills. The brothers also created the vault of the heavens with his skull and his brain became the fleecy clouds.

The ancient Chinese concept of creation was one of chaos. The early universe had the shape of a chicken's egg. After 18000 years, the egg broke releasing its contents. The lighter part of it (the Yang) formed the sky and the heavier part (the Yin) condensed to form the Earth. Also released from the egg was a giant named Pan-Ku. Pan-Ku's body magically grew about three meters each day, eventually dividing the sky from the Earth. After another 18000 years, Pan-Ku died. From his head were created the Sun and the Moon, from his blood the rivers and seas, from his breath the wind and from his voice the thunder.

In the Western Hemisphere, the Mayan priesthood devised a mythology in which the Earth was flat with four corners. At each corner, there was a jaguar of a different color that supported the sky. The Mayans further believed that the universe was divided into layers, with each layer containing only one type of celestial body. Heaven had thirteen layers, each with its own god. The underworld had nine layers, with nine corresponding lords of the night. When the Mayans observed the Sun and the Moon setting, they believed that each passed through the Earth after it disappeared below the horizon.

In the fifth century BC, Pythagoras and his followers were familiar with the concept of a spherical, rotating Earth; Oenopides calculated the tilt of the Earth's axis to be 24°, and Meton developed the 19-year Metonic cycle of eclipses. Half a century later, Aristotle wrote about every aspect of science. He considered the daily motions of the Sun, Moon, planets and stars and stated that the Earth was stationary, and in the center of the universe. Democritus stated that the Moon was similar to the Earth. Heraclides suggested that Mercury and Venus orbit the Sun, which in turn orbits the Earth. Eudoxus of Knidos used a series of rotating spheres to account for the motion of the Sun.

In the third century BC, Aristarchus of Samos taught that the Sun, not the Earth, was the center of the planetary system. Aristarchus also became the first to calculate distances to the Sun and Moon. Eratosthenes of Alexandria calculated the circumference of the Earth. In the following century, Hipparchus discovered the phenomenon of precession and, using a lunar eclipse, derived a reasonably accurate distance to the Moon. He also offset the (still circular) orbit of the Earth to account for the seasonal variations observed at that time.

Several hundred years later, in the second century AD, Claudius Ptolemaeus (Ptolemy) wrote the *Almagest* (*Syntaxis*), a 13-volume treatise on astronomy. In this work, Ptolemy presents his theory of epicyclic motion for the planets. For better or worse, this work remained the standard exposition on astronomy for more than a thousand years.

In the period between Ptolemy and the Renaissance, many writers have been tempted into stating that nothing happened. In fact, while it is true that no "breakthroughs" were announced, astronomical observations were still being conducted, most notably in China. Planetary observations, occultations, eclipses, and records of the appearances of comets and guest stars (supernovae) continued unabated throughout Europe's Dark Ages.

Correspondingly, observing techniques and instrumentation were slowly being improved. Arabic astronomy – with the help of such notables as Al-Battani and Ulugh Beg – produced several voluminous sets of astronomical and mathematical tables, the most memorable being those commissioned by Alfonso X of Castile, since referred to as the Alfonsine Tables.

Important concepts

The Earth is not in the center of the solar system!

Niklas Koppernigk, better known as Nicolas Copernicus, was a Renaissance man. Indeed, it may be fair to say that he helped define the Renaissance by writing, in this author's opinion, the most influential science book of all time, *De Revolutionibus Orbium Coelestium*. The manuscript of this work was entrusted to the Austrian astronomer Georg Joachim von Lauchen, also known as Rheticus. In 1540, Rheticus became the first to publish an account of the Copernican theory in *Narratio prima de libris revolutionum Copernici*. He oversaw the publication of *De Revolutionibus Orbium Coelestium* in Nuremberg, in 1543.

Ptolemy. (Michael
E. Bakich
collection)

Ptolemy. (Michael E. Bakich collection)

Years before this monumental work was published, Copernicus had circulated to a few friends and astronomers a short commentary outlining his views. This appeared about 1512. In this work he makes a number of assumptions. Among them, Copernicus claims that there is no true center for all celestial bodies, that all the planets circle the Sun and that the Earth is among them, that the heavenly motions are in fact due to the motions of the Earth, its rotation about its axis and its revolution about the Sun, and that the retrograde motions of the planets can also be accounted for by the Earth's motion.

Copernicus argued that the apparent annual motion of the Sun about the Earth could be equally well represented by a motion of the Earth about the Sun, and the rotation of the celestial sphere could be accounted for by assuming that the Earth rotates about a fixed axis while the celestial sphere is stationary. Ptolemy taught that all heavy matter tends to move to the center of the world and remain at rest there. If the Earth rotated, everything would be torn apart. Copernicus countered this argument by stating that if such motion would tear the Earth apart then the even faster motion of the celestial sphere, due to its greater size, would be even more damaging to it.

Copernicus's arguments were elegant and persuasive. But it was the philosophical, rather than the scientific, implication of his theory that was groundbreaking. The main point of Copernicus's idea is that the Earth is not something special, but merely one of several planets in revolution about the Sun. This idea, that the Earth is "typical" rather than "special" has been a driving force in astronomy ever since Copernicus. Even today, astronomers

The world system according to Ptolemy. Zahn, Johann, *Specula physico-mathematico-historica notabilium ac mirabilium sciendorum*, Nuremberg, 1696. (Linda Hall Library)

assume that the same physical laws that are at work in our cosmic neighborhood can be applied similarly throughout the universe.

The formation of the Moon

In the early part of the nineteenth century it was suggested that the origin of the Moon was a simple one: when the Earth formed, the Moon formed nearby. This is called the "common condensation" theory or sometimes the "double planet" theory. The main problem with this theory is the very different densities of the Earth and Moon. The Earth's density is $5.515 \, \text{g/cm}^3$, while that of the Moon is only $3.34 \, \text{g/cm}^3$. If the Moon formed next to the Earth, and at the same time, its density should match almost exactly.

In 1878, British astronomer Sir George Howard Darwin proposed that the Moon was once part of the Earth. According to Darwin, this was a time near the formation of the Earth, when the planet was still molten. Because of the rapid rotation of the Earth, a large piece was broken off, eventually becoming the Moon. This is known as the "fission" theory. Darwin proposed the Pacific Ocean basin as the place from which the Moon was ejected. Unfortunately, no satisfactory mechanism has ever been suggested to account for such an event. Also, if the Moon were thrown off of a spinning Earth, it would orbit in the plane of the Earth's equator, and not be inclined (5.145°) as it is.

A third possibility, proposed at the start of the twentieth century, is known as the "capture" theory. As the name implies, this theory states that a separate astronomical object encountered our planet at a very close distance and was captured by the gravitational field of the Earth. This sounds plausible,

The world system according to Tycho Brahe. Zahn, Johann, *Specula physico-mathematico-historica notabilium ac mirabilium sciendorum*, Nuremberg, 1696. (Linda Hall Library)

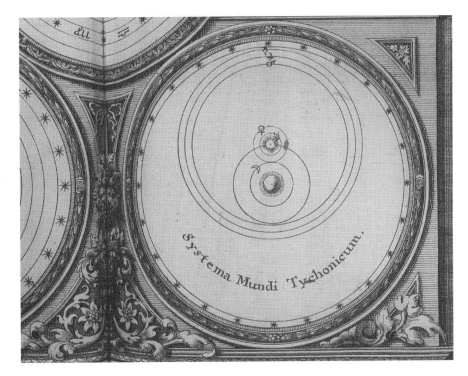

The world system according to Nicolas Copernicus. Zahn, Johann, *Specula physico-mathematico-historica notabilium ac mirabilium sciendorum*, Nuremberg, 1696. (Linda Hall Library)

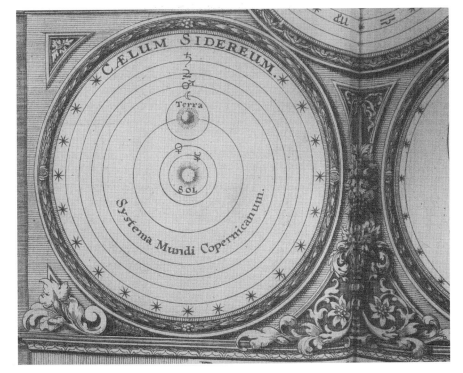

The Moon.
Hevelius,
Johannes,
Selenographia:
sive, Lunae
descriptio, Gdansk,
1647. (Linda Hall
Library)

but the mechanics of the situation almost demand the presence of a third body. In such a chance encounter, the interaction of three objects results in one of them (the Moon) being slowed to an orbital speed.

New light was shed on this problem when the US Apollo astronauts returned samples of rock and soil gathered from the surface of the Moon during six missions from 1969–1972. Suddenly, not only was the density of the Moon known but also its chemical composition. It was learned that the Moon has a similar composition to the Earth's crust. Both have approximately the same proportions of silicon, magnesium, manganese and iron. The Moon has far smaller proportions of volatiles, but higher proportions of non-volatiles such as aluminum and titanium. In the minds of most astronomers, the differences in chemical composition, along with the Moon's lack of an iron core, ruled out both the common condensation theory and the fission theory.

In the mid-1970s, US astronomers William K. Hartmann and Donald R. Davis proposed an alternative theory of the Moon's formation. According to this new hypothesis, the Moon formed from debris blasted out of the Earth by the impact of a Mars-sized body. The great age of lunar rocks and the absence of any impact feature on Earth indicate that this event must have occurred during the Earth's own formation, some 4.5 billion years ago. The

The Lunar Roving Vehicle made collecting specimens easier. (NASA)

"Big Splat" theory, as it is called, answers many questions that have been raised.

Such an impact would vaporize elements with low melting points and disperse them. Since only the crust and outer layer of the Earth's mantle would be blasted out, Earth's iron core would remain intact. This explains the low density and low iron content of lunar material. The differences in composition between the Earth and the Moon can be accounted for by the additional material of the impacting body. Even the Earth's 23.45° tilt can be explained: it was knocked over by the blast!

In 1997, Sigeru Ida, Robin M. Canup and Glen R. Stewart presented data regarding the impact. They found that the cloud of vaporized rock ejected by the blast flattens into a disk after a few months. According to their calculations, about 2/3 of this material falls back to Earth. To produce a satellite the size of the Moon, they propose a minimum impactor mass of 2 to 3 times that of Mars. They point out, however, that such an impact would leave the Earth–Moon system with twice as much angular momentum as it has today and offered no mechanism to reduce this initial angular momentum.

How many moons does Earth possess?

In 1846, Frédéric Petit, director of the observatory of Toulouse, stated that a second moon of the Earth had been discovered. It had been seen during the early evening of 21 March 1846. Petit calculated an orbit that was elliptical, with a period of 2 hours 44 minutes 59 seconds, an apogee at 3570 km above the Earth's surface and perigee at just 11.4 km above the Earth's surface.

52) "We came in peace …". Neil Armstrong's footprint on the surface of the Moon. (NASA)

Astronomers generally ignored reports such as this, and the idea would have been forgotten had a young French writer, Jules Verne, not read about it. In his novel *From the Earth to the Moon* Verne lets a small object pass close to the traveler's space capsule, causing the capsule to travel around the Moon instead of smashing into it.

Jules Verne was read by millions of people, and made Petit's second moon known all over the world. Amateur astronomers jumped at the opportunity for fame – anybody discovering this second moon would have his name inscribed in the annals of science.

W. H. Pickering did not look for the Petit object, but he did carry out a search for a secondary moon – a satellite of our Moon ("On a photographic search for a satellite of the Moon," *Popular Astronomy*, 1903). The result was negative and Pickering concluded that any satellite of our Moon must be smaller than about 3 m.

Pickering's article on the possibility of a tiny second moon of Earth, "A Meteoritic Satellite," appeared in *Popular Astronomy* in 1922 and caused another short flurry among amateur astronomers, since it contained a virtual request: "A 9-cm telescope with a low-power eyepiece would be the likeliest means to find it. It is an opportunity for the amateur." But again, all searches remained fruitless.

There have been other proposals for additional natural satellites of the Earth. In 1898 Georg Waltemath from Hamburg claimed to have discovered not only a second moon but a whole system of midget moons. Waltemath worked out an orbit for the largest of these moons. "Sometimes," he said, "it shines at night like the Sun." Public interest was aroused when Waltemath predicted his second moon would pass in front of the Sun on the 2nd, 3rd or 4th of February 1898. Not surprisingly this was not observed.

A very early printed Moon photograph. The area shown is approximately 600 km × 350 km. Guillemin, Amedée, *The Heavens*, London, 1866. (Michael E. Bakich collection)

Lunar earthrise.
(NASA)

Contemporary astronomers were irritated by the public's persistent interest in the new moon. But astrologers caught on – in 1918 the British astrologer Sepharial named this moon Lilith. He considered it to be black enough to be invisible most of the time, being visible only close to opposition or when in transit across the solar disk. Sepharial constructed an ephemeris of Lilith, based on several of Waltemath's claimed observations. He considered Lilith to have about the same mass as the Moon, apparently happily unaware that any such satellite would, even if invisible, show its existence by perturbing the motion of the Earth. And even to this day, "the dark moon" Lilith is used by some astrologers in their horoscopes.

From time to time other "additional moons" were reported by observers. The German astronomical magazine *Die Sterne* reported that a German amateur astronomer named W. Spill had observed a second moon cross our first moon's disk on May 24 1926.

The Earth *can* have a very near satellite for a short time. Meteoroids passing the Earth and skimming through the upper atmosphere can lose enough velocity to go into a satellite orbit around the Earth. But since they pass the upper atmosphere at each perigee, they will not last long, maybe only one or two, possibly a hundred revolutions (about 150 hours). There are some indications that such "ephemeral satellites" have been seen; it is even possible that Petit's observers did see one.

People are still proposing additional natural satellites of the Earth. Between 1966 and 1969 John Bargby, a US scientist, claimed to have observed at least ten small natural satellites of the Earth, visible only in a telescope. Bargby calculated orbits for all the objects. He considered them to be fragments of a larger body which broke up in December 1955. He based much of his theory

The Lunar Apennines, Archimedes Crater, etc. The area shown is approximately 650 km across. Nasmyth, James & Carpenter, James, *The Moon: Considered as a Planet, a World, and a Satellite*, London, 1874. (Michael E. Bakich collection)

of the suggested satellites on supposed perturbations of artificial satellites. Bargby used artificial satellite data from the *Goddard Satellite Situation Report*, unaware that the values in this publication are only approximate and sometimes grossly in error and can therefore not be used for any precise scientific analysis. In addition, from Bargby's own claimed observations it can be deduced that when at perigee Bargby's satellites ought to be visible at first magnitude and thus be easily visible to the naked eye, yet no-one else has seen them.

Interesting facts

Earth is the densest planet in the solar system, with an average density of 5.515 g/cm^3.

The Moon has 1940 named features, of which 1545 are craters.

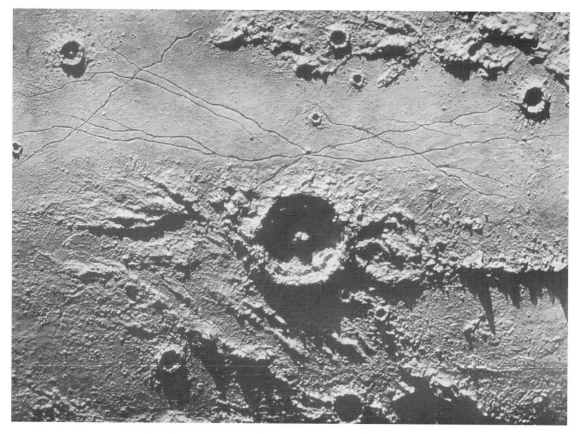

Triesnecker Crater (approximately 23 km in diameter). Nasmyth, James & Carpenter, James, *The Moon: Considered as a Planet, a World, and a Satellite*, London, 1874. (Michael E. Bakich collection)

The Earth from space. (NASA)

The composition of seawater, apart from the hydrogen and oxygen is as follows:

Chloride (Cl) 55.04%   Magnesium (Mg) 3.69%
Sulfate (SO_4) 7.69%   Potassium (K) 1.10%
Calcium (Ca) 1.16%   Bromide (Br) 0.10%
Bicarbonate (HCO_3) 0.41%   Fluorine (F) 0.003%
Strontium (Sr) 0.04%   Trace elements 0.157%
Sodium (Na) 30.61%

Earth's four oceans contain 97.2% of the water on our planet. The Antarctic ice sheet, which can be up to 4800 m thick, contains 70% of the Earth's fresh water.

The largest crater on the Moon, Hertzsprung, has a diameter of 591 km. It is the second largest crater in the solar system (next to Beethoven on Mercury) and the largest crater on any satellite.

If all the ice on Earth melted, the sea level would rise by 100 m.

According to the Belgian mathematician Jean Meeus, the extreme distances between the centers of the Earth and the Moon are minimum = 356371 km and maximum = 406720 km. Meeus evaluated ten centuries' worth of data from 1500–2500 AD.

The Earth's polar diameter is 44 km less than its equatorial diameter.

The Full Moon, though bright, is only 1/400000 as bright as the Sun. If the entire sky were covered with Full Moons, we would receive only about 1/5 the illumination of the Sun on a bright day.

The total surface area of the Earth is 509509000 km², of which 70.7% is water and 29.3% is land.

The First and Last Quarter Moons are only about 10% as bright as the Full Moon.

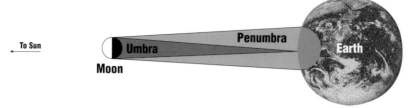

Geometry of a solar eclipse; not to scale. (Holley Bakich)

Future eclipses *Total solar eclipses*

Date	Time (ET)	Maximum duration
11 Aug 1999	11^h	02^m23^s
21 Jun 2001	12^h	04^m56^s
4 Dec 2002	08^h	02^m04^s
23 Nov 2003	23^h	01^m57^s
8 Apr 2005	21^h	00^m42^s
29 Mar 2006	10^h	04^m07^s
1 Aug 2008	10^h	02^m28^s
22 Jul 2009	03^h	06^m40^s
11 Jul 2010	20^h	05^m20^s
13 Nov 2012	22^h	04^m02^s
3 Nov 2013	13^h	01^m40^s
20 Mar 2015	10^h	02^m47^s
9 Mar 2016	02^h	04^m10^s
21 Aug 2017	18^h	02^m40^s
2 Jul 2019	19^h	04^m32^s
14 Dec 2020	16^h	02^m10^s

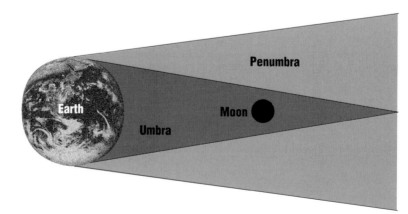

Geometry of a lunar eclipse; not to scale. (Holley Bakich)

Total lunar eclipses

Date	Time (ET)	Duration of totality
21 Jan 2000	04^{h}45^m	01^{h}16^m
16 Jul 2000	13^{h}57^m	01^{h}43^m
9 Jan 2001	20^{h}22^m	01^{h}00^m
16 May 2003	03^{h}41^m	00^{h}52^m
9 Nov 2003	01^{h}20^m	00^{h}22^m
4 May 2004	20^{h}32^m	01^{h}16^m
28 Oct 2004	03^{h}05^m	01^{h}20^m
3 Mar 2007	23^{h}22^m	01^{h}14^m
28 Aug 2007	10^{h}38^m	01^{h}30^m
21 Feb 2008	03^{h}27^m	00^{h}50^m
21 Dec 2010	08^{h}18^m	01^{h}12^m
15 Jun 2011	20^{h}13^m	01^{h}40^m
10 Dec 2011	14^{h}33^m	00^{h}50^m
15 Apr 2015	07^{h}48^m	01^{h}18^m
8 Oct 2014	10^{h}55^m	00^{h}58^m
28 Sep 2015	02^{h}48^m	01^{h}12^m
31 Jan 2018	13^{h}31^m	01^{h}16^m
27 Jul 2018	20^{h}23^m	01^{h}42^m
21 Jan 2019	05^{h}13^m	01^{h}02^m

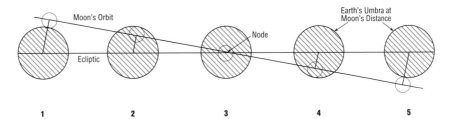

Geometry of lunar eclipses, continued. If the position of the Moon is between 2 and 4, a total eclipse will occur. If the Moon is between 1 and 2 or between 4 and 5, a partial eclipse will occur. This diagram may be expanded to show penumbral eclipses as well. (Holley Bakich)

The Moon. Galilei, Galileo, *Sidereus Nuncius*, Venice, 1610. (Linda Hall Library)

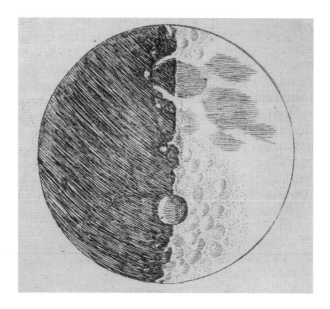

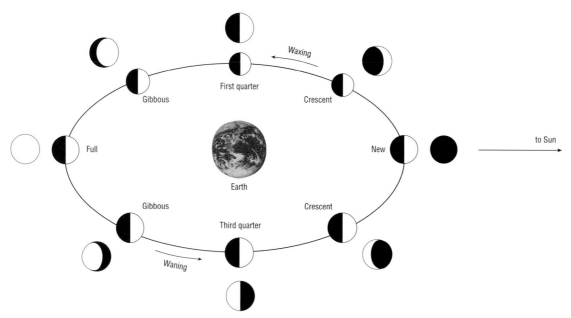

Lunar phases and the Moon's appearance from Earth. (Holley Bakich)

Satellite

	Size (km)	Mass (kg)	Density (g/cm³)	Orbital period	Eccentricity	Inclination	Distance from planet (km)
Moon	3476	7.15×10^{22}	3.340	$27^{d}07^{h}43.7^{m}$	0.0549	5.145°	3.844×10^{5}

An early
explanation of
lunar phases.
Ozanam, Jacque,
*Cours de
mathématique –
Nouv. éd. rév. et
corr.*, Paris, 1699
(Linda Hall Library)

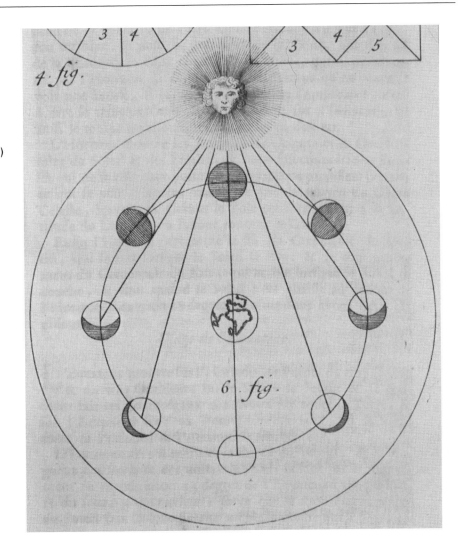

Historical timeline

c. 22 000 BC	Meteor Crater is in northern Arizona, US. The crater is 1265 m in diameter and 174 m deep.
22 October 4004 BC	The date of Creation; at least, according to Archbishop James Ussher who worked out a chronology based upon the descendants of Adam, as listed in the Bible; and, for those who are interested, the time of the Creation was 8 p.m.
22 October 2137 BC	According to the *Historical Document of Ancient China*, a total solar eclipse was observed.
800 BC	The Chinese *Book of Changes* records the earliest observation of a sunspot.
45 BC	Julius Caesar mandated the introduction of the Julian calendar.

29 BC	The Chinese began making regular observations of sunspots by looking through slices of jade.
19 June 325	The Council of Nicea was convened by the Emperor Constantine; among other items, the rules for determining the dates of Easter were implemented (the first Sunday following the first Full Moon on or after the vernal equinox); Easter remains the only western holiday still determined by astronomical means.
22 March 837	The first sighting of Halley's Comet, during its apparition of this year, occurred. This is generally regarded as the most spectacular appearance of this cosmic wanderer. Reports state that it became as bright as the Full Moon with a tail stretching halfway across the sky.
1054	The Chinese sighted a "guest star" in the constellation Taurus, visible in daylight for 23 days and at night with the naked eye for two years. Today, when astronomers point their telescopes to the area, they see the remnants of a supernova – the Crab Nebula.
2 April 1066	Halley's Comet became visible in the constellation of Pegasus; it disappeared after 67 days. This is the apparition made famous by its depiction on the Bayeux Tapestry.
19 February 1473	Nicolas Copernicus was born in Torun, Poland.
1505–1514	Italian painter and scientist Leonardo da Vinci made sketches which show surface features on the Moon; these are the earliest such drawings which have survived.
1543	One of the greatest of scientific works, *De Revolutionibus Orbium Coelestium*, was published shortly before the death of its author, Nicolas Copernicus, on 24 May.
15 February 1564	Italian astronomer Galileo Galilei was born.
11 November 1572	The Danish astronomer Tycho Brahe made his first observation of "Tycho's Star," a supernova in the constellation of Cassiopeia; it was first seen by Chinese observers on 6 November.
24 February 1582	To correct the observed seasonal irregularities, Pope Gregory XIII issued a papal bull decreeing that 4 October would be followed by 15 October; thus marking the dawn of the Gregorian calendar.
9 October 1604	The first sighting of "Kepler's Star," a supernova in the constellation of Ophiuchus, was made in China; Kepler first observed it a few days later.
September 1608	Dutch spectacle-maker Jan Lippershey constructed a telescope.

Jan Lippershey.
(Michael E. Bakich
collection)

16 May 1609	Galileo first heard a report of a Dutch telescope.
30 November 1609	Galileo's first celestial observations were made with a telescope of his construction.
25 August 1835	The first in a series of articles which would come to be known as "The Great Moon Hoax" appeared in the *New York Sun*. These articles, which were total fabrication, supposedly describe Sir John Herschel's observations of various types of lunar life, while he was observing the Moon from South Africa.
30 June 1908	At approximately 07:17 local time, a gigantic explosion occurred over the Tunguska region of Central Siberia, devastating thousands of square kilometers of forest; it is believed that the event was the collision between the Earth and a small comet nucleus, as no trace of the impactor has ever been found.
16 March 1926	The first liquid fuel rocket was launched by the US scientist Robert Goddard; the rocket reached an altitude of 56 m.

THE SUN.

NEW YORK, TUESDAY MORNING, AUGUST 25, 1835. [Price One Cent.

GREAT ASTRONOMICAL DISCOVERIES,
LATELY MADE
BY SIR JOHN HERSCHEL, L.L.D., F.R.S. & c.
At the Cape of Good Hope

The "Great Moon Hoax" fooled a lot of people in August 1835. (Michael E. Bakich collection)

12 February 1947	At 10:40 local time, an iron–nickel meteor exploded over the Sikote-Alin Mountains north of Vladivostok, Russia. The bolide was widely viewed and became brighter than the Sun. The initial mass of the meteoroid is estimated at 9072 kg. The largest recovered meteorite weighed 1745 kg.
24 February 1949	The first rocket into outer space was a US Bumper-WAC, launched from White Sands, NM; it reached an altitude of 395 km.
20 September 1951	The first successful space flight for living creatures occurred when a US Aerobee rocket carried a monkey and 11 mice to an altitude of 350 km; they were later recovered alive.
4 October 1957	The Space Age officially began in the (then) USSR with the successful launch of Sputnik ("fellow traveler"), the first artificial satellite.

2 January 1959	USSR probe Luna 1 launched; Luna 1 made the first lunar flyby. It found the solar wind and is now in a solar orbit.
3 March 1959	US probe Pioneer 4 launched for a distant flyby of the Moon; probe is now orbiting the Sun.
12 September 1959	USSR probe Luna 2 launched; Luna 2 was the first spacecraft to impact on the surface of the Moon, on September 14 1959.
4 October 1959	USSR probe Luna 3 launched; far-side flyby; encountered the Moon on October 7 1959 and returned the first image of the Moon's hidden side. Space probe is now in a decayed Earth–Moon orbit.
4 April 1960	The first weather satellite was launched, the US Tiros 1; sent back cloud photographs. (Tiros stands for "Television Infra-Red Orbital Satellite.")
12 August 1960	The first communications satellite, Echo 1, was launched; Echo 1 was a passive satellite, essentially a 30.5-m diameter balloon which reflected signals sent from Earth back down to the surface.
4 October 1960	The first active communications satellite, Courier 1B, was launched.
12 April 1961	The first human in space was the Soviet cosmonaut Yuri Alekseyevich Gagarin; he made one orbit of the Earth and safely returned the same day.
23 April 1962	US probe Ranger 4 launched; first US lunar impact; struck Moon's surface on 25 April 1962; no data returned.
18 October 1962	US probe Ranger 5 launched; Ranger 5 was to be a lander but became a flyby because of a spacecraft failure; now in solar orbit.
2 April 1963	USSR probe Luna 4 launched; Luna 4 was intended to be a lunar lander but missed the Moon; now in Earth–Moon orbit.
30 January 1964	US probe Ranger 6 launched; cameras failed; probe impacted the surface of the Moon.
28 July 1964	US probe Ranger 7 launched; arrived 31 July 1964, sent pictures back at a close range, and impacted the Moon.
17 February 1965	US probe Ranger 8 launched; arrived at the Moon on 20 February 1965; sent back 7100 high-resolution pictures until it impacted in Mare Tranquillitatis.
21 March 1965	US probe Ranger 9 launched; sent back 5800 high-resolution photographs before impacting on the floor of the Alphonsus Crater on 24 March 1965.
9 May 1965	USSR probe Luna 5 launched; soft-lander failed and it impacted the Moon.
8 June 1965	USSR probe Luna 6 launched; missed the Moon by 160 900 km; now in orbit around the Sun.

The Moon. Wright, Thomas, *An Original Theory or New Hypothesis of the Universe*, London, 1750. (Linda Hall Library)

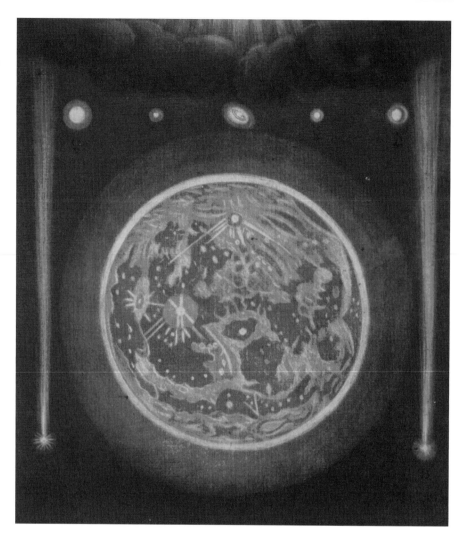

18 July 1965	USSR probe Zond 3 launched; returned 25 pictures of the lunar far side on its way to Mars; now in solar orbit.
4 October 1965	USSR probe Luna 7 launched; failed and impacted the Moon.
3 December 1965	USSR probe Luna 8 launched; failed and impacted the Moon.
31 January 1966	USSR probe Luna 9 launched; first soft-landing on the Moon; returned 27 photographs from the surface.
31 March 1966	USSR probe Luna 10 launched; orbiter; returned data for two months; still in lunar orbit.
30 May 1966	US probe Surveyor 1 launched; first US soft-landing on the lunar surface, in Oceanus Procellarum; 11 150 photographs transmitted.

10 August 1966	US probe Lunar Orbiter 1 launched; first US craft to circle the Moon; after sending 211 high-resolution photographs, it was crashed into the lunar surface to avoid conflict with Lunar Orbiter 2.
24 August 1966	USSR probe Luna 11 launched; orbiter still in lunar orbit.
20 September 1966	US probe Surveyor 2 failed and crash-landed in Oceanus Procellarum.
22 October 1966	USSR probe Luna 12 launched; orbiter sent back high-resolution photographs; still in lunar orbit.
6 November 1966	US probe Lunar Orbiter 2 launched; circled the Moon and photographed 13 possible Apollo landing sites; crashed onto lunar surface to avoid Lunar Orbiter 3.
21 December 1966	USSR probe Luna 13 launched; soft-landing on lunar surface; sent back photographs and tested soil.
5 February 1967	US probe Lunar Orbiter 3 launched; orbited the Moon, photographed the far side for potential Apollo landing sites, then impacted on command.
17 April 1967	US probe Surveyor 3 launched; soft-landed on the floor of Oceanus Procellarum; sent back 6315 photographs; the astronauts from Apollo 12 visited this craft during their time on the Moon.
4 May 1967	US probe Lunar Orbiter 4 launched; sent into a polar orbit and transmitted first photographs of Moon's south pole; measured lunar gravity and radiation; crashed into lunar surface to avoid Lunar Orbiter 5.
14 July 1967	US probe Surveyor 4 launched; made soft-landing in Sinus Medii, then radio contact was lost.
19 July 1967	US probe Explorer 35 launched; orbited Moon and acquired field and particle data.
1 August 1967	US probe Lunar Orbiter 5 launched; orbited the Moon at a polar inclination; took high-resolution pictures of many important sites; impacted on command.
8 September 1967	US probe Surveyor 5 launched; landed in Mare Tranquillitatis; analyzed soil and sent back 18006 photographs.
7 November 1967	US probe Surveyor 6 launched; soft-landed in Sinus Medii; analyzed soil; sent back 30000 photographs; took off from the lunar surface.
7 January 1968	US probe Surveyor 7 launched; soft-landed near Tycho Crater; analyzed soil; sent back 21000 photographs.
7 April 1968	USSR probe Luna 14 launched; orbited the Moon and measured gravitational field; now in a lunar–solar orbit.

14 September 1968	USSR probe Zond 5 launched; biosatellite containing turtles, worms, flies, bacteria, spiderwort plant and seeds of wheat, pine and barley; flew around Moon to within 1951 km; took photographs of far side; returned to Earth and splashed down in the Indian Ocean.
10 November 1968	USSR probe Zond 6 launched; biosatellite also containing micrometeorite and cosmic ray detectors; flew around Moon to within 2250 km; returned to Earth, landing in USSR.
21 December 1968	Apollo 8 launched. Crew: Frank Borman, James A. Lovell, Jr, William Anders; first staffed lunar fly around and Earth return; the astronauts made 10 orbits of the Moon on Christmas Eve; returned to Earth 27 December 1968.
18 May 1969	Apollo 10 launched. Crew: Thomas Stafford, Eugene A. Cernan, John W. Young; lunar fly around and Earth return; Stafford and Cernan tested the Lunar Module, separating it from the Command and Service Module and descended to within 15 240 m of the lunar surface; returned to Earth 26 May 1969.
13 July 1969	USSR probe Luna 15 launched; unsuccessful sample return attempt; crashed during landing.
16 July 1969	Apollo 11 launched. Crew: Neil A. Armstrong, Edwin E. Aldrin, Jr, Michael Collins; first staffed lunar landing, which took place on July 20 1969; landing site was Mare Tranquillitatis at latitude 0°67′ N and longitude 23°49′ E; Armstrong and Aldrin collected 21.7 kg of soil and rock samples and deployed experiments; returned to Earth 24 July 1969.
8 August 1969	USSR probe Zond 7 launched; flew around Moon and returned; took color photographs.
14 November 1969	Apollo 12 launched. Crew: Charles Conrad, Jr, Alan L. Bean, Richard F. Gordon, Jr; manned lunar landing which took place on November 19 1969; landing site was Oceanus Procellarum at latitude 03°12′ S and longitude 23°23′ W; Conrad and Bean retrieved portions of Surveyor 3, including the camera; samples amounting to 34.4 kg were returned; astronauts also deployed the Apollo lunar surface experiment package (ALSEP), an automated research station which was also deployed by all subsequent lunar crews; returned to Earth 24 November 1969.
11 April 1970	Apollo 13 launched. Crew: James A. Lovell, Jr, Fred W. Haise, Jr, John L. Swigert, Jr; aborted mission; during the translunar coast an explosion destroyed both power and propulsion systems of the Command and Service Module; Lunar Module was used as a lifeboat for the astronauts; returned to Earth 17 April 1970.

12 September 1970	USSR probe Luna 16 launched; landed on 20 September 1970 at Mare Fecunditatis located at latitude 0°41′ S and longitude 56°18′ E; a return vehicle brought 100 g of lunar samples to Earth.
20 October 1970	USSR probe Zond 8 launched; orbited Moon within 1101 km; first television images of Earth from a distance of 64 374 km.
10 November 1970	USSR probe Luna 17 launched; made lunar landing with an automated Lunokhod 1 rover vehicle.
31 January 1971	Apollo 14 launched. Crew: Alan B. Shepard, Jr, Edgar D. Mitchell, Stuart A. Roosa; Shepard and Mitchell landed on the Moon on 5 February 1971, in the Fra Mauro highlands, located at latitude 03°40′ S and longitude 17°28′ E; they collected 42.9 kg of lunar samples; returned to Earth 8 February 1971.
26 July 1971	Apollo 15 launched. Crew: David R. Scott, James B. Irwin, Alfred M. Worden; Scott and Irwin landed on the Moon on 30 July 1971 at a site named Hadley-Apennine, latitude 26°06′ N and longitude 03°39′ E; they collected 76.8 kg of samples; first use of Lunar Roving Vehicle; Command and Service Module was the first to carry orbital sensors and to release a sub-satellite into lunar orbit; Worden performed the first deep spacewalk to retrieve film from the service module; returned to Earth 7 August 1971.
2 September 1971	USSR probe Luna 18 launched; unsuccessful sample return attempt; crashed during landing.
28 September 1971	USSR probe Luna 19 launched; orbited Moon and transmitted photographs; now in lunar orbit.
14 February 1972	USSR probe Luna 20 launched; landed on 21 February 1972 at Apollonius highlands located at latitude 03°32′ N and longitude 56°33′ E; 30 g of lunar samples returned to Earth.
16 April 1972	Apollo 16 launched. Crew: John W. Young, Charles M. Duke, Jr, Thomas K. Mattingly II; Young and Duke landed on 21 April 1972, at the Descartes Crater located at latitude 09°00′ N and longitude 15°31′ E; they deployed instruments, drove the lunar rover, and collected 94.7 kg of samples during a 71-hour surface stay; returned to Earth 27 April 1972.
7 December 1972	Apollo 17 launched. Crew: Eugene A. Cernan, Harrison H. Schmitt, Ronald B. Evans; Cernan and Schmitt landed on the Moon on 12 December 1972; landing site was Taurus-Littrow at latitude 20°10′ N and longitude 30°46′ E; they returned 110.5 kg of rock and soil samples; astronauts covered 30.5 km in the lunar rover during a 75-hour stay; returned to Earth 19 December 1972.

8 January 1973	USSR probe Luna 21 launched; lunar landing with an automated Lunokhod 2 rover vehicle.
29 May 1974	USSR probe Luna 22 launched; orbiter; entered lunar orbit; transmitted photographs to Earth.
28 October 1974	USSR probe Luna 23 launched; soft-landing on lunar surface; sample return to Earth failed.
9 August 1976	USSR probe Luna 24 launched; landing site was Mare Crisium at latitude 12°45′ N and longitude 60°12′ E; 170 g of samples were brought back to Earth.
24 January 1990	Japan probe Muses-A launched; first non-US or USSR probe to reach Moon; orbited Moon; failed to send back data.
8 December 1990	US–European space probe Galileo, on its way to Jupiter, flew by the Earth–Moon system for a gravity boost.
25 January 1994	US probe Clementine launched; a new design using lightweight structure and propellant systems; 90 days (6 February–5 May 1994) in lunar orbit; four cameras mapped the surface of the Moon at 125–250 meters/pixel resolution; gathered altimeter data making it possible to generate the first lunar topographic map.
17 February 1998	Voyager 1 overtakes Pioneer 10 as the furthest human-made object from Earth.
11 August 1999	A total solar eclipse will occur; mid-eclipse is approximately 11^h ET; duration of totality will be 02^m23^s.
21 June 2001	A total solar eclipse will occur; mid-eclipse is approximately 12^h ET; duration of totality will be 04^m56^s.
4 December 2002	A total solar eclipse will occur; mid-eclipse is approximately 08^h ET; duration of totality will be 02^m04^s.
23 November 2003	A total solar eclipse will occur; mid-eclipse is approximately 23^h ET; duration of totality will be 01^m57^s.
8 April 2005	An annular-total solar eclipse will occur; mid-eclipse is approximately 21^h ET; duration of totality will be 00^m42^s.
29 March 2006	A total solar eclipse will occur; mid-eclipse is approximately 10^h ET; duration of totality will be 04^m07^s.
1 August 2008	A total solar eclipse will occur; mid-eclipse is approximately 10^h ET; duration of totality will be 02^m28^s.
22 July 2009	A total solar eclipse will occur; mid-eclipse is approximately 03^h ET; duration of totality will be 06^m40^s.
11 July 2010	A total solar eclipse will occur; mid-eclipse is approximately 20^h ET; duration of totality will be 05^m20^s.
13 Nov 2012	A total solar eclipse will occur; mid-eclipse is approximately 22^h ET; duration of totality will be 04^m02^s.

Mars

Physical data

Size   6794.4 km*

Mass 6.421×10^{23} kg

Escape velocity   5.02 km/s (18072 km/hr)

Temperature range   Minimum   Average   Maximum
$-140\,°C$ $-63\,°C$ $20\,°C$

Oblateness   0.00519

Surface gravity   3.72 m/s^2

Volume 1.643×10^{11} km^3
15.1% that of Earth

Magnetic field strength and orientation   Mars has little or no magnetic field due to the fact that, unlike the Earth, the core of the planet is not liquid metal. Mars, like the Moon, lacks sufficient mass to produce the pressure (and, therefore, the temperature) associated with maintaining a liquid core.

Albedo (visual geometric albedo)   0.15

Density (water $= 1$)   3.94 g/cm^3

Solar irradiance   595 watts/m^2

Atmospheric pressure   699 to 912 pascals

Composition of atmosphere

Carbon dioxide (CO_2)	$>95.3\%$
Molecular nitrogen (N_2)	$>2.6\%$
Argon (Ar)	1.6%
Molecular oxygen (O_2)	0.13%
Carbon monoxide (CO)	0.07%
Water vapor (H_2O)	0.03%
Neon (Ne)	0.00025%
Krypton (Kr)	0.00003%
Xenon (Xe)	0.000008%
Ozone (O_3)	0.000003%

Maximum wind speeds   100 km/hr

* Polar diameter 6759 km.

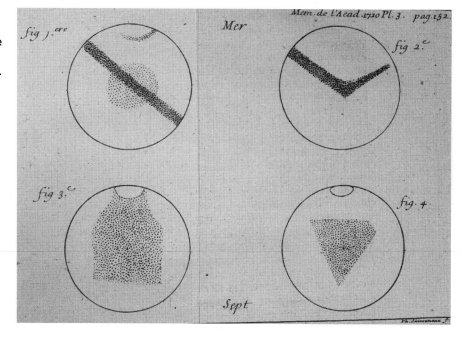

The first sketches of Mars to depict known features; the last of the four shows Syrtis Major. Maraldi, Giacomo Filippo, "Observations sur les taches de Mars" *Histoire de l'Académie Royale des Sciences*, Année 1720:144–53, Paris, 1722. (Linda Hall Library)

Outstanding cloud and surface features

Cloud features

The Martian atmosphere can be divided into a troposphere and a stratosphere. The troposphere is quite variable. During the Martian day, the troposphere can extend to a height of 10 km, while at night, or over the colder polar regions, the troposphere essentially disappears.

All Martian clouds are temporary phenomena. High thin clouds composed of H_2O ice may form in mountainous regions. In addition, in low-lying areas of the planet fog is possible in the hours just prior to sunrise. Also, a stratospheric haze of CO_2 may form when crystals of dry ice condense at high altitudes.

Though not truly clouds, dust storms have been observed from time to time on Mars. These storms can be quite severe, enveloping the entire surface area of the planet for up to several months.

Surface features

The land area of Mars is approximately equivalent to the dry land area of Earth. Due to the smaller size of the planet, the thin Martian atmosphere and the lack of erosion, large surface features on Mars tend to be more pronounced than those on Earth. Here is a brief list of the largest of the 1345 named features on Mars:

Feature	Diameter (km)
Alba Fossa	2077
Arabia Terra	6000
Arcadia Planitia	3052
Capri Chasma	1498
Cassini Crater	415
Claritas Fossa	2033
Copernicus Crater	292
Daedalia Planum	2477
Elysium Planitia	3899
Herschel Crater	304
Huygens Crater	456
Icaria Fossae	2153
Ius Chasma	1003
Kasei Valles	2222
Kovalsky Crater	299
Lacus Planum	2900
Lycus Sulcus	1639
Nepenthes Mensa	1704
Newton Crater	287
Nilokeras Scopulus	1064
Noachis Terra	3500
Novi Cavus	535
Oenotria Scopulus	1438
Olympus Mons	550
Olympus Rupes	1819
Promethei Rupes	1491
Scandia Colles	1232
Schiaparelli Crater	461
Schroeter Crater	337
Sirenum Fossae	2712
Tharsis Montes	2105
Tikhonravov Crater	390
Utopia Planitia	3276
Valles Marineris	4128
Xanthe Terra	3074

The most prominent surface features on Mars are the polar ice caps. Each of the ice caps can be subdivided into "seasonal" and "residual" caps. The southern residual ice cap is approximately 350 km in diameter and is composed of frozen CO_2. The northern residual ice cap is composed of H_2O ice, and has a diameter of 1000 km. Both seasonal ice caps are composed of frozen CO_2, which condenses directly from the Martian atmosphere when the temperature is below $-123\,°C$. In the northern hemisphere, where the winters are more severe, the extent of the ice cap may reach a latitude of $45°$, but in the

Numerous sketches of Mars. Herschel, William, "On the remarkable appearances at the polar regions of the planet Mars" *Philosophical Transactions*, **74**:233–73, London, 1784. (Linda Hall Library)

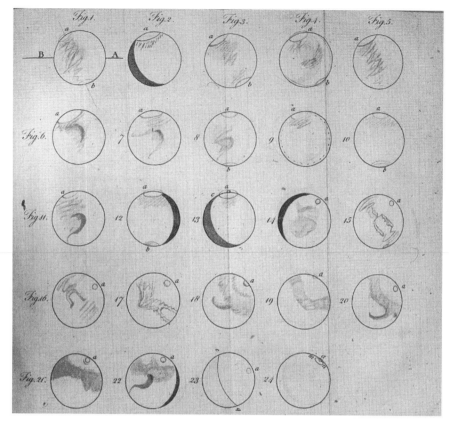

southern hemisphere the extent of the seasonal ice cap never passes below a latitude of about 55°.

Orbital data

Period of rotation   1.025957 days
24.6240 hours
1^{d}0^{h}37.4^m

Period of revolution (sidereal orbital period)   1.8809 years
686.98 days
1^{y}320^{d}18.2^h

Synodic period   779^{d}22^{h}33.6^m

Equatorial velocity of rotation   866.9 km/hr

Velocity of revolution   24.13 km/s (86868 km/hr)

Distance from Sun   Average   1.5237 AU
227940000 km
Maximum   1.6661 AU
249251000 km
Minimum   1.3811 AU
206615600 km

Distance from Earth	Maximum	2.68 AU
		401 355 980 km
	Minimum	0.36 AU
		54 510 620 km

Apparent size of Sun (average)   0.35°

Apparent brightness of Sun $m_{vis} = -25.8$

Orbital eccentricity   0.0934

Orbital inclination   1.850°

Inclination of equator to orbit   25.19°

Observational data

Maximum angular distance from Sun   180°

Brilliancy at opposition	Maximum	−2.9
	Minimum	−1.0

Angular size*	Maximum	25.16″
	Minimum	3.5″

Early ideas

In Assyria, the planet Mars was the special sign of Nergal, often called "Shedder of Blood," a god of death, misfortune and disaster. To the Norsemen, Mars was Tyr, or Tiu, the one-handed god of war. From this god we derive the name of the third day of the week, "Tiu's day." The planet represented the god of war to the ancient Greeks as well. There, it was known as Ares. The Romans made both the mineral hematite and iron sacred to Mars, their red god of war, and the stone or metal was often used as an amulet during battle.

To the Etruscans, Mars was not originally a god of war. His earliest image was that of a sacrificed fertility god, Maris, worshiped at an ancient shrine in northern Latium. He joined with the goddess Marica and their union produced Latinus, the legendary ancestor of all Latin tribes.

More than 2500 years ago, the Babylonians were making regular observations of Mars, which they named Salbatani. A Babylonian text which research has shown dates from January 523 BC, records the observation of a heliacal setting of Mars, in the western part of the constellation Gemini. The ancient Chinese, who called Mars Huo xing, used the planet astrologically and the following quote indicates that they watched its motions carefully: "When Mars is retrograding in the station Ying-she, the ministers conspire and the soldiers revolt."

Aristotle taught that the Earth was the center of the universe and the Sun, Moon and planets revolved around it, describing perfectly circular orbits. But simple observation showed that there were irregularities. For instance, the

---

* This measurement is the apparent angular diameter of Mars, measured in seconds of arc, as seen from Earth.

Mars. *Annals of the Astronomical Observatory of Harvard College*, **2**, part 8, Cambridge, MA, 1876. (Michael E. Bakich collection)

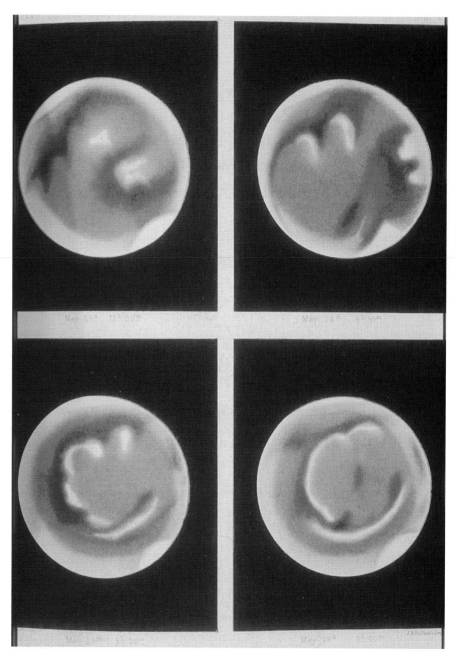

brightnesses of the planets changed and each planet did not always move through the starry background at the same velocity. Most hostile to Aristotle's theory was that the direction of motion of the planets changed on a regular basis. This was most pronounced in the case of Mars. The retrograde motion of the red planet seemed intensely damaging to a theory which suggested circular motions. To account for this, Aristotle and Eudoxus each had systems

The retrograde motion of Mars. As the faster-moving Earth passes Mars at opposition, Mars appears to reverse direction. Since the orbit of Mars is inclined with respect to the ecliptic, the retrograde segment of the orbit does not overlay the direct segment. (Holley Bakich)

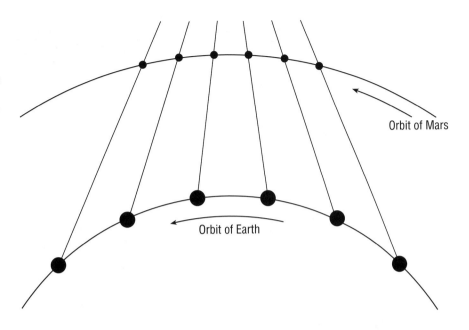

View From Earth

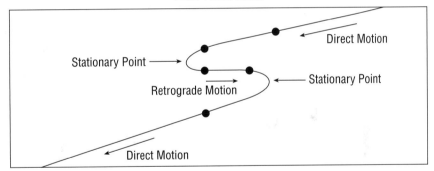

of rotating, nested spheres upon the surfaces of which rested Mars and the other planets. This was an incredibly complex system. Eudoxus utilized 27 spheres and Aristotle's system demanded 55, 22 of which were counter-rotating.

Two centuries later, a somewhat simpler system was championed by Hipparchus. This was the theory of epicycles. Planets were assumed to describe a circle (called an epicycle) around a point which was part of a larger circle (called a deferent) which itself was centered on the Earth. It is important to note that all motions were circular. It would have been heresy to suggest otherwise.

Again, two centuries passed. In the Library of Alexandria, Claudius Ptolemaeus, also known as Ptolemy, made numerous observations of Mars with which to refine the epicycle theory. Ptolemy introduced inclinations between

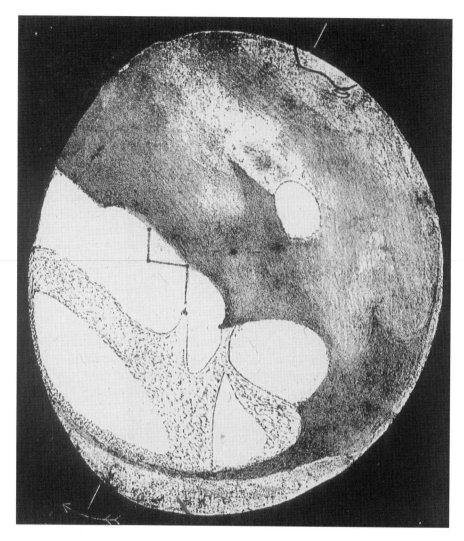

the deferents and the ecliptic and between the epicycles and the deferents. He added deferents upon deferents and the whole system became extremely complex. However, to a rough degree of accuracy it did predict the positions and motions of Mars and the other planets unlike any theory before it.

Mars was the focal point for Johannes Kepler's three laws of planetary motion. Kepler worked for more than five years to obtain an orbit of Mars whereby prediction agreed with observation. He once suggested that the Sun had a repulsive force which was variable with distance, causing a type of epicyclic orbit of the planet. But even this did not remove the error between what was expected and what was seen. Eventually, Kepler realized that his difficulties were caused by his unwillingness to forego circular orbits. As soon as he applied an elliptical orbit to Mars, the errors vanished.

In 1672 Jean Richer was in the French colony of Cayenne making observations of Mars. The previous year Jean-Dominique Cassini was at the Paris

Images of Mars by Hooke (left) and Cassini (right). Hooke, Robert and Cassini, Jean-Dominique, "The particulars of those observations of the planet Mars" *Philosophical Transactions*, **1**:239–47, London, 1666. (Linda Hall Library)

Observatory taking similar measurements. When their two values were compared, the first (reasonably) accurate value of the parallax of Mars was deduced and, from that, its distance. Making observations of Mars during favorable oppositions (those when Mars is near perihelion) was one of two ways that astronomers attempted to deduce the distance scale of the solar system. The other was by using transits of Venus.

The rotational rate of Mars was refined by the Dutch astronomer Friedrich Kaiser. During the opposition of 1862, he made a number of drawings, and even constructed a globe. Then he compared markings on his maps to similar characteristics on maps drawn by Christiaan Huygens in 1666, and Robert Hooke in 1667. The value Kaiser obtained for the rotational period of Mars differs from the presently accepted value by only 0.1 second.

During the exceptional opposition of 1877, the Scottish astronomer David Gill traveled to Ascension Island to measure the parallax of Mars. Gill was trying a relatively new method – measuring the position of Mars against the background of stars twice in the same night. The first observation is made early in the evening when Mars is in the east and then another is made early in the morning when Mars is in the west. Thus, a lone observer could measure the east–west component of Mars's parallax avoiding the errors inherent in using a second observer a great distance away.

It was also during the 1877 opposition of Mars that the Italian astronomer Giovanni Virginio Schiaparelli caused such a stir. But more on that – and Percival Lowell – later.

Important concepts

Kepler and the Laws of Planetary Motion

Following the philosophy of the great Aristotle, the Greek astronomer Ptolemy, imagined the Earth in the center of an unchanging universe, with Sun, Moon and planets orbiting it in perfect circular motion. The Polish astronomer Nicolas Copernicus replaced the Earth with the Sun. The Danish astronomer Tycho Brahe insisted the Earth was the center, but imagined that the planets orbited the Sun, which orbited the immobile Earth. The German

Mars. Ozanam,
Jacque, *Cours de
mathématique –
Nouv. éd. rév. et
corr.*, Paris, 1699.
(Linda Hall Library)

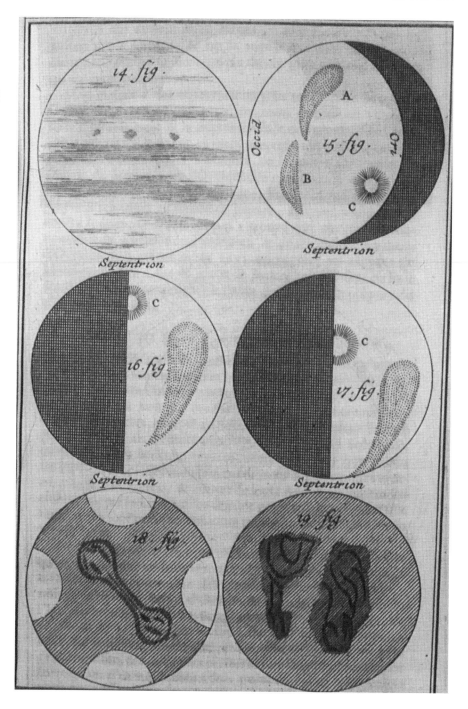

Tycho Brahe.
(Michael E. Bakich
collection)

theoretician Johannes Kepler added a slight wrinkle to Copernicus's theory, and, in so doing, unlocked the mystery of planetary motion.

Tycho and Kepler knew of each other and even corresponded. In 1601 Tycho arranged to have King Rudolf II appoint Kepler to the post of Imperial Mathematician. Kepler's assignment was to assist Tycho in the observations of the planets and the creation of new planetary tables. Kepler and Tycho did not get along well. Tycho tended to secrecy, apparently mistrusting Kepler, fearing that his bright young assistant might eclipse him as the premiere astronomer of his day. He therefore let Kepler see only part of his observational data.

Tycho set Kepler the task of understanding the orbit of the planet Mars, which was particularly troublesome. It is believed that part of the motivation for giving the Mars problem to Kepler was that it *was* difficult, and Tycho hoped it would occupy Kepler while Tycho worked on his theory of the solar system. In a supreme irony, it was precisely the Martian data that allowed Kepler to formulate the correct laws of planetary motion, thus achieving a place in the development of astronomy far surpassing that of Tycho.

Soon after Kepler set to work, Tycho died. Kepler obtained Tycho's data despite the attempts by Brahe's family to keep the data from him in the hope of monetary gain. There is some evidence that Kepler obtained the data by less than legal means; it is fortunate for the development of modern astronomy that he was successful. Unlike Tycho, Kepler believed firmly in the Copernican system. In retrospect, the reason that the orbit of Mars was particularly difficult was that Copernicus had correctly placed the Sun at the center of the solar system, but had erred in assuming the orbits of the planets to be circles. Thus, in the Copernican theory epicycles were still required to explain the details of planetary motion.

It fell to Kepler to provide the final piece of the puzzle. He had once stated that, if given all of Tycho's data, he could calculate a planetary orbit in a week. After a long struggle, Kepler was forced finally to the realization that the orbits of the planets were not the circles demanded by Aristotle and assumed implicitly by Copernicus, but were instead the "flattened circles" that geometers call ellipses.

The irony noted above lies in the realization that the difficulties with the Martian orbit derive precisely from the fact that the orbit of Mars was the most elliptical of the planets for which Brahe had extensive data. Thus Brahe had unwittingly given Kepler the very part of his data that would allow Kepler to eventually formulate the correct theory of the solar system and thereby to banish Brahe's own theory!

Here then are the three Laws of Planetary Motion, as revealed by Johannes Kepler:

Kepler's first law:
I. The orbits of the planets are ellipses, with the Sun at one focus of the ellipse.

The Sun is not at the center of the ellipse, but is instead at one focus (there is nothing at the other focus of the ellipse). The planet then follows the ellipse in its orbit, which means that the Earth–Sun distance is constantly changing as the planet revolves.

Kepler's second law:
II. A line joining the planet to the Sun sweeps out equal areas in equal amounts of time.

The result of this law is the realization that each planet in its orbit moves faster when it is nearer the Sun. Thus, a planet executes elliptical motion with

Johannes Kepler.
(Michael E. Bakich
collection)

constantly changing angular speed as it moves about its orbit. Hence, by Kepler's second law, the planet moves fastest when it is near perihelion and slowest when it is near aphelion.

Kepler's third law:
III. The ratio of the squares of the orbital periods for two planets is equal to the ratio of the cubes of their semimajor axes.

The result of this law is the realization that, with regard to their position relative to the Sun, closer planets move faster than further planets. Mercury moves faster than Venus which moves faster than the Earth, etc.

177

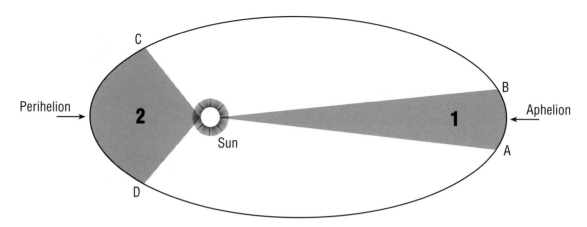

Kepler's second law. For the hypothetical planet, assume A to B takes one month and C to D takes one month. According to Kepler's second law, the area of 1 equals the area of 2. (Holley Bakich)

Percival Lowell

No book with a chapter on Mars would be complete without some discussion of Percival Lowell (1855–1916). Lowell was a tireless observer of Mars and although his scientific conclusions concerning life on the red planet were erroneous, there is no debate that Lowell – almost single-handedly – brought the study of Mars and the excitement of astronomical discovery to the public at large. Lowell was an engaging lecturer, a popular writer, the founder of a major observatory and the major proponent of a trans-Neptunian planet. With all these credentials, why do modern astronomers view his theories with disdain? Rather than launching into a detailed narrative, I thought it best to allow Lowell to speak for himself, that the reader may judge. The following excerpts are from the first edition of the first of Lowell's astronomical works, *Mars* (Houghton, Mifflin and Co., Boston and New York, 1895).

Consider an introductory statement:

. . . So much for matter. As for that manifestation of it known as mind, modesty, if not intelligence, forbids the thought that we are sole thinkers in all we see. Indeed, we seldom stop in our locally engrossing pursuits to realize how small the part we play in the universal drama. [p. 4]

Lowell was the first great proponent of locating observatories in areas of good seeing. Despite the fact that he used this argument to bolster his claim of seeing canals, he was correct in his science:

The reason that so few astronomers have as yet succeeded in seeing these lines is to be found in our own atmosphere . . . A moderately good air is essential to their detection; and unfortunately the locations of most of our observatories preclude this prerequisite . . . It is not simply a question of a clear air, but of a steady one. To detect fine detail, the atmospheric strata must be as evenly disposed as possible. [pp. 138–9]

178

Mars. Unpublished photograph of a 5½″ (14 cm) Lowell observatory globe, 1894. (Linda Hall Library)

Lowell was not shy about interpreting his observations:

When we put all these phenomena together – the presence of the spots at the junctions of the canals, their strangely systematic shapes, their seasonal darkening, and, last but not least, the resemblance of the great continental regions of Mars to the deserts of the earth – a solution of their character suggests itself at once; to wit, that they are oases in the midst of that desert, and oases not wholly innocent of design . . . [pp. 185–6]

About his detractors, Lowell had this to say:

It is interesting to recall, in connection with this incredulity about the canals, that precisely the same thing happened in the case of the discovery of Jupiter's satellites and with Huygens' explanation of Saturn's ring. We are apt to imagine that our age of the world has a monopoly of skepticism. But this is a mistake. The spirit that denies has always been abroad; only in early days he was reputed to be the devil. [p. 149]

Amazingly, the answer was within his reach:

Our senses are our avenues of approach from the outer world. Messages from them are therefore usually and rightly attributed to stimuli from without. But it is possible for these messages to be tampered with at any stage of their journey. It is even possible for them to be started in some other part of the brain, travel down to the lower centres and be sent up from them to the higher ones, indistinguishable from *bona fide* messages from without. Bright points in the sky or a blow on the head will equally

179

Percival Lowell observing Mars. (Michael E. Bakich collection)

cause one to see stars. In the first case the eyes were duly affected from without; in the second, the nerves were tapped to the same effect in mid-route; but in each case the subsequent current travels to the higher centres apparently as authentic the one as the other. [p. 159]

Lowell's conclusions, however, left little room for opposing views:

To review, now, the chain of reasoning by which we have been led to regard it probable that upon the surface of Mars we see the effects of local intelligence . . . it is evident that what we see, and call by ellipsis the canal, is not really the canal at all, but the strip of fertilized land bordering it . . . A planet may in a very real sense be said to have life of its own . . . It is born, has its fiery youth . . . Now, in the special case of Mars, we have before us the spectacle of a world relatively well on in years . . . Certainly what we see hints at the existence of beings who are in advance of, not behind us, in the journey of life . . . [pp. 201–9]

According to Lowell, it was our collective fear of something different that held humanity (more specifically, his peers in the astronomical community) back from accepting what he knew to be true:

Its [life on Mars] strangeness is a purely subjective phenomenon, arising from the instinctive reluctance of man to admit the possibility of peers . . . To admit into his conception of the cosmos other finite minds as factors has in it something of the weird. Any hypothesis to explain the facts, no matter how improbable or even palpably absurd it be, is better than this . . . It is simply an instinct like any other, the projection of the instinct of self-preservation. [pp. 209–10]

Indeed, it would be ironic – some might say fitting – if the book's closing statement turns out to be prophetic:

If astronomy teaches anything, it teaches that man is but a detail in the evolution of the universe, and that resemblant though diverse details are inevitably to be expected

180

Mars, various
views. Proctor,
Richard Anthony,
*Old and New
Astronomy*,
London, 1892.
(Michael E. Bakich
collection)

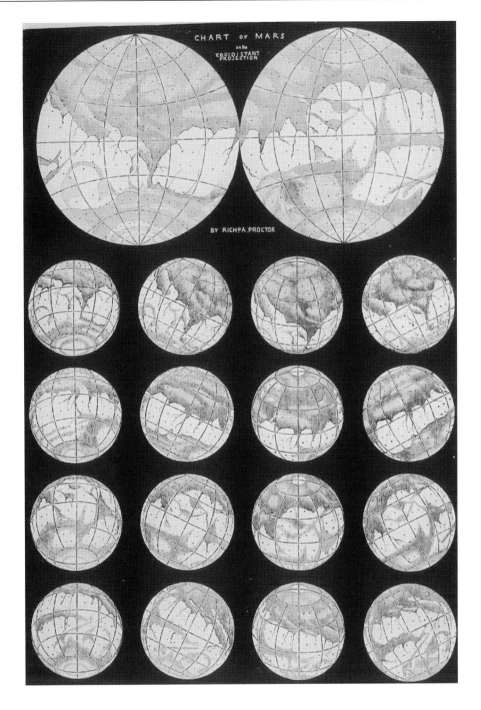

The man who
frightened
America, Orson
Welles, 1938.
(Michael E. Bakich
collection)

in the host of orbs around him. He learns that, though he will probably never find his double anywhere, he is destined to discover any number of cousins scattered throughout space. [p. 212]

Lowell was not alone

Ladies and gentlemen, we interrupt our program of dance music to bring you a special bulletin from the intercontinental radio news. At twenty minutes before eight, central time, Professor Farrel of the Mt Jennings Observatory, Chicago, Illinois, reports observing several explosions of incandescent gas occurring at regular intervals on the planet Mars . . . Ladies and gentlemen, here is the latest bulletin . . . It is reported that at 8:50 p.m., a huge flaming object, believed to be a meteorite, fell on a farm in the neighborhood of Grover's Mill, New Jersey . . . Someone crawling out or something I can see peering out of that black hole – two luminous disks, are the eyes, or it might be a face; might be almost anything. Good heavens! Something's wriggling out of the shadow like a gray snake. Now it's another one and another one and another one. Wait, I can see the thing's body now. It's large, large as a bear. It glistens like wet leather. That face, it – ladies and gentlemen, it's indescribable. I can hardly force myself to keep looking at it, it's so awful . . . Wait a minute, something's happening. Humped shape is rising out of the pit. I can make out a small beam of light against a mirror. What's there? There's a jet of flame springing from that mirror. It leaps right at the advancing men. It strikes them head on! Lord, they're turning into flames! . . .

On Halloween evening, 1938, US broadcaster Orson Welles produced one of the most frightening radio programs in history. Based upon H. G. Wells' classic science fiction novel of 1898, the "War of the Worlds" broadcast seized untold numbers of Americans in a momentary wave of mass hysteria. Despite a clear explanation of the nature of the program at the beginning of the broadcast, thousands of listeners were actually convinced that an interplanetary war was in progress.

The "War of the Worlds" broadcast brought to a zenith the interest and fear associated with the question of life on Mars. But the stage had been set and the drama had been building for many years.

In the eighteenth century, the philosopher Immanuel Kant speculated on the mental capacities of the inhabitants of other worlds. He thought that the beings of the inferior planets were too material to be reasonable, and were probably not even responsible for their actions. He ranked the human life of Earth and Mars as a happy moral medium, neither absolutely coarse nor absolutely spiritual:

These two planets are placed in the middle of our planetary system, so that we may suppose, without improbability, that their inhabitants possess an average condition, in their constitutions as well as in their morals, between the two extremes. [*Metaphysical Foundations of Natural Science*, 1786]

Kant also described the perfection and the happiness which the inhabitants of the superior planets enjoy. This was typical of philosophical assumptions regarding life on Mars for the 200 years following the invention of the telescope.

In the 1780s, Sir William Herschel suggested that Mars experienced seasons like those of the Earth and that it possessed "a considerable but moderate atmosphere." To these conservative statements, however, he added that the inhabitants of Mars "probably enjoy a situation similar in many respects to our own." Herschel, like other astronomers of his time, thought that all the planets might be inhabited, and not only believed in life on the Moon at an early point in his career, but later stated that the Sun was "richly stored with inhabitants."

During the nineteenth century, as observational techniques continued to improve, many astronomers were turning their attention to Mars; and for good reason. Mars, at first glance, looks very Earth-like. It is the nearest planet whose surface we can see. There are vast polar ice caps, seasonally changing patterns and dust storms, a 24-hour day and an atmosphere.

One of the most noted observers of that period was Giovanni Virginio Schiaparelli who made many careful observations of Mars from Milan. During a close approach of Mars to the Earth in 1877, Schiaparelli observed and plotted an intricate network of single and double straight lines criss-crossing the bright areas of the planet. These were published on a map in 1890. He was not the first to see these linear markings, but his use of the term "canali," an Italian word which means channels or grooves, contributed to many myths surrounding these features, myths that were to linger until the middle of the twentieth century. This was due in part to a loose translation of the term "canali," which altered it to "canals," a word that implied intelligent design.

One of the first popularizers of the canals on Mars was the French astronomer and author Camille Flammarion. He wrote at length on the subject of life on Mars and suggested different ways to communicate with its

Giovanni Virginio
Schiaparelli.
(Michael E. Bakich
collection)

inhabitants. Flammarion's writings had vast popular appeal and, in at least one instance, influenced the decision of a wealthy, but somewhat eccentric, patron of the sciences that communication with intelligence on Mars was a foregone conclusion.

The individual, Madame Guzman, bequeathed quite a large sum of money to the French Academy of Sciences to be presented to the first person who established communication with life on another world – that is, except for Mars. In her own words:

I bequeath to the Academy of Sciences of the Institute of France, 100 000 francs for founding a prize which carries the name of my son, Pierre Guzman. This prize is to be given, without exception to nationality, to the first person who finds the means of communicating with a star – by this I meant to say a signal to a star and a received response to that signal, exclude the planet Mars because it is sufficiently well-known. [*Archives of the French Academy of Sciences*]

The years following the initial "discovery" of the canals on Mars saw a number of manuscripts produced, supposedly detailing what was occurring on that planet. A few of these were scientific in nature, but most were speculative and outrageous. In 1894, John Jacob Astor published *Journey in Other*

The image that started it all. Mars. Schiaparelli, Giovanni Virginio, *Il pianeta marte*, Milan, 1893. (Linda Hall Library)

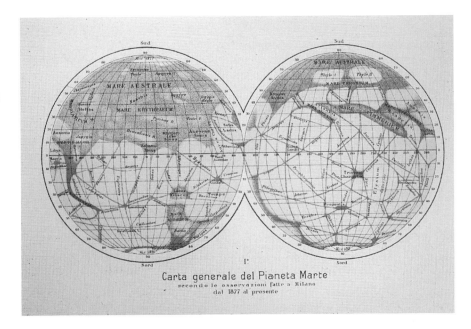

Carta generale del Pianeta Marte
secondo le osservazioni fatte a Milano
dal 1877 al presente

Worlds, and in 1899, Ellsworth Douglass flew to Mars in a cigar-shaped spaceship (*Cassell's*). Douglass had Martians living in a biblical setting, with the hero landing in the Martian equivalent of Egypt. Early in the twentieth century, C. E. Housden, in *Riddle of Mars*, worked out the form and power that the supposed pumping stations required to irrigate Mars as the Martians appeared to be doing.

Edgar Rice Burroughs, the creator of Tarzan, was perhaps the most famous popularizer of Mars. His first "Martian" novel, *A Princess of Mars* was published in 1912. Burroughs took Percival Lowell's speculations about Mars and fictionally populated the dying planet with green, six-legged beasts of burden, cliffs of solid gold, men of all colors (including green) and, of course, immense canals. Throughout this entire period it is amazing that the only aspect of Mars that was considered certain was the existence of the canals!

Filmmakers were not about to let pass such fertile science fiction soil. Some of the Mars films with which this writer is familiar: *Rocketship X-M* (1950); *Flight to Mars* (1951); *Red Planet Mars* (1952); *Invaders from Mars* (1953, remade in 1986); *The War of the Worlds* (1953); *The Angry Red Planet* (1960); *A Martian in Paris* (1961); *Mars Needs Women* (1966); *Planet of Blood* (1966); *Total Recall* (1990); *Mars Attacks* (1997).

Percival Lowell may have been the first, but he was not alone.

Martian meteorites

On 6 August 1996, NASA Administrator Daniel S. Goldin made the following statement:

A Princess of Mars by Edgar Rice Burroughs. (Michael E. Bakich collection)

NASA has made a startling discovery that points to the possibility that a primitive form of microscopic life may have existed on Mars more than three billion years ago. The research is based on a sophisticated examination of an ancient Martian meteorite that landed on Earth some 13000 years ago.

The evidence is exciting, even compelling, but not conclusive. It is a discovery that demands further scientific investigation. NASA is ready to assist the process of rigorous scientific investigation and lively scientific debate that will follow this discovery.

I want everyone to understand that we are not talking about "little green men." These are extremely small, single-cell structures that somewhat resemble bacteria on Earth. There is no evidence or suggestion that any higher life form ever existed on Mars.

The NASA research team at the Johnson Space Center, Houston, TX, and at Stanford University, Palo Alto, CA, had found evidence that strongly suggested primitive life may have existed on Mars more than 3.6 billion years ago.

Among the highlights of the team's two-year research was the detection of an apparently unique pattern of organic molecules. Several unusual mineral phases that are known products of primitive microscopic organisms on Earth

One of the all-time best science fiction books. (Michael E. Bakich collection)

RAY BRADBURY **THE MARTIAN CHRONICLES**

were also found. Structures that could be microscopic fossils seem to support all of this. The team's excitement stemmed from the fact that each of these pieces of evidence was found within a few hundred thousandths of a centimeter of one another.

The igneous rock in the 1.9-kg, potato-sized meteorite has been age-dated to about 4.5 billion years, the period when the planet Mars formed. The rock is believed to have originated underneath the Martian surface and to have been extensively fractured by impacts as meteorites bombarded the planets in the early inner solar system. Between 3.6 billion and 4 billion years ago, a time when it is generally thought that the planet was warmer and wetter, water is believed to have penetrated fractures in the subsurface rock, possibly forming an underground water system.

Since the water was saturated with carbon dioxide from the Martian atmosphere, carbonate minerals were deposited in the fractures. The team's findings indicate living organisms may also have assisted in the formation of the carbonate, and some remains of the microscopic organisms may have become fossilized, in a fashion similar to the formation of fossils in limestone on Earth. Then, 16 million years ago, a huge comet or asteroid struck Mars, ejecting a piece of the rock from its subsurface location with enough force to escape the planet (5.02 km/s). For millions of years, the chunk of rock floated through space. It encountered Earth's atmosphere 13 000 years ago and fell in Antarctica as a meteorite.

It is in the tiny globs of carbonate that the researchers found a number of features that can be interpreted as suggesting past life. Stanford researchers found easily detectable amounts of organic molecules called polycyclic aromatic hydrocarbons concentrated in the vicinity of the carbonate. This finding appears consistent with the proposition that they are a result of the fossilization process. In addition, the unique composition of the meteorite's polycyclic aromatic hydrocarbons is consistent with what the scientists expect from the fossilization of very primitive microorganisms. On Earth, polycyclic aromatic hydrocarbons virtually always occur in thousands of forms, but in the meteorite they are dominated by only about a half-dozen different compounds.

The team found unusual compounds – iron sulfides and magnetite – that can be produced by anaerobic bacteria and other microscopic organisms on Earth. The compounds were found in locations directly associated with the fossil-like structures and carbonate globules in the meteorite. The carbonate also contained tiny grains of magnetite that are almost identical to magnetic fossil remnants often left by certain bacteria found on Earth.

The largest of the possible fossils are less than 1/100 the diameter of a human hair, and most are about 1/1000 the diameter of a human hair. Some are egg-shaped while others are tubular. In appearance and size, the structures are strikingly similar to microscopic fossils of the tiniest bacteria found on Earth.

The meteorite, designated ALH84001, was found in 1984 in Allan Hills ice field, Antarctica, by an annual expedition of the National Science Foundation's Antarctic Meteorite Program. Its possible Martian origin was not recognized until 1993. It is one of only 12 meteorites identified so far that match the unique Martian chemistry measured by the Viking spacecraft that landed on Mars in 1976. ALH84001 is by far the oldest of the 12 Martian meteorites, more than three times as old as any other.

Of course the primary question asked of the investigators was whether or not the material found within the meteorite could have been contaminated with organic material from Earth. The formation of the carbonate or fossils by living organisms while the meteorite was in the Antarctic was deemed unlikely for several reasons. The carbonate was age-dated and found to be 3.6 billion years old, and the organic molecules were first detected well within the ancient carbonate. In addition, the team analyzed representative samples of other meteorites from Antarctica and found no evidence of fossil-like structures, organic molecules or possible biologically produced compounds and minerals similar to those in the ALH84001 meteorite.

The composition and location of organic molecules found in the meteorite also appeared to confirm that the evidence of possible life was extraterrestrial. None were found in the meteorite's exterior crust, but the concentration increased in the meteorite's interior to levels higher than ever found in Antarctica. Higher concentrations of polycyclic aromatic hydrocarbons would have been likely to have been found on the exterior of the meteorite, decreasing toward the interior, if the organic molecules were the result of contamination of the meteorite on Earth.

Martian meteorites

Name	Classification	Mass (kg)[a]	Find/Fall[b]	Year
Chassigny	C-dunite (olivine)	4.00	fall	1815
Shergotty	S-basalt	4.00	fall	1865
Nakhla	N-clinopyroxenite	40.00	fall	1911
Lafayette	N-clinopyroxenite	0.80	find	1931
Gov. Valadares	N-clinopyroxenite	0.16	find	1958
Zagami	S-basalt	18.00	fall	1962
ALHA77005	S-lherzolite	0.48	find-A	1978
EETA79001	S-basalt	7.90	find-A	1980
LEW88516	S-lherzolite	0.013	find-A	1991
ALH84001	orthopyroxenite	1.90	find-A	1993
QUE94201	S-basalt	0.012	find-A	1995
Y793605	S-lherzolite	0.018	find-A	1995

Notes:
[a] Values given are the most accurate available at the time of writing.
[b] Find-A indicates that the meteorite was a recent find in Antarctica.

Early "observations" concerning the moons of Mars

The first astronomer to state that Mars had moons was the father of the Laws of Planetary Motion, Johannes Kepler, in 1610. When trying to solve Galileo's problem referring to Saturn's rings (which Galileo thought to be two moons close to the planet), Kepler believed that Galileo had found moons of Mars instead.

In 1643, the Capuchin monk Anton Maria Shyrl claimed to have observed two moons in orbit around Mars. We now know that such a view would be

Mars, various.
Zahn, Johann,
*Specula physico-
mathematico-
historica
notabilium ac
mirabilium
sciendorum*,
Nuremberg, 1696.
(Linda Hall Library)

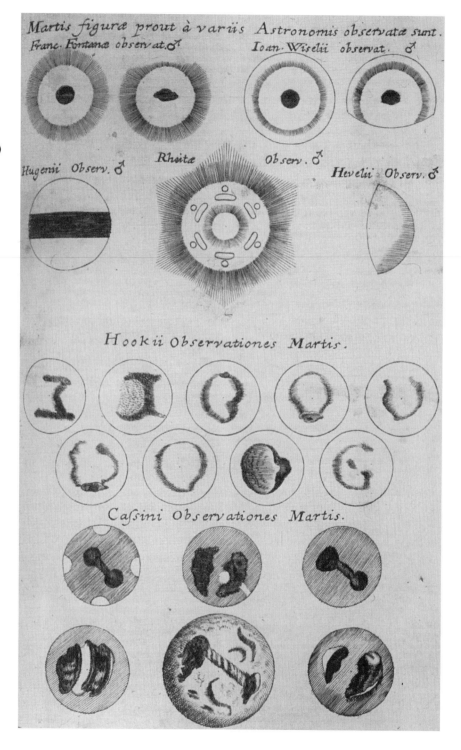

impossible with the telescopes of his time. In all likelihood, Shyrl had observed a star in the same field of view as Mars.

Of all the conjectures related to the satellites of Mars, certainly the most famous took place in the eighteenth century. One hundred and fifty years prior to their actual discovery, Jonathan Swift wrote *Gulliver's Travels* (1726). He mentions two small moons orbiting Mars. These satellites were known to the astronomers of Lilliput, having been observed by them for some time. Swift gives their periods of revolution as 21 and 10 hours. Swift's imagined moons were again mentioned in 1750, by Voltaire in his novel *Micromegas*. This was the story of a giant from the star Sirius who visits our solar system.

Twenty years after *Gulliver's Travels*, in 1747, a German military captain named Kindermann claimed to have observed a satellite of Mars three years earlier, on 10 July 1744. Kindermann had worked out an orbit for this "moon" and stated that its period of revolution around Mars was 59 hours 50 minutes and 6 seconds.

In 1877, Asaph Hall, an astronomer working at the US Naval Observatory, finally discovered the two small moons of Mars. He was given the honor of naming these satellites and, in deference to the mythology concerning the god of war, chose the names Phobos (fear) and Deimos (dread). The orbital periods of these two satellites are 7 hours 39 minutes and 30 hours 18 minutes, remarkably close to the periods imagined by Jonathan Swift 150 years earlier.

Interesting facts

Under favorable nighttime conditions, the light from Mars may cast a visible shadow.

The only known incident of a meteorite killing a mammal occurred in 1911, in Nakhla, Egypt. The fall killed a dog and was witnessed by the dog's owner. The meteorite has been shown to be of Martian origin.

From Mars, the Sun appears 44% as bright as from Earth.

During the nineteenth century, supposed inhabitants of Mars were often referred to as "Martials."

Mars has a total of 1345 named features, of which 845 are craters. The largest crater on Mars, named Schiaparelli, is 461 km in diameter.

The Martian rotational period (approximately 24^h37^m) is referred to as a sol. Scientists at the Jet Propulsion Laboratory in Pasadena, CA, use this term to differentiate between it and an Earth "day."

The Earth is 9.3 times as massive as Mars.

On 27 August 2003, Mars will reach its maximum brightness, shining almost as bright as magnitude −3.0. The most recent approach of Mars to this magnitude was on 22 August 1924. More recently, the red planet has twice shone at magnitude −2.9, on 7 September 1956 and on 12 August 1971. Because of changing surface features on Mars (most notably the reflective polar ice caps), opposition brightness may differ by as much as 0.3 magnitude from prediction.

The second smallest naturally occurring satellite in the solar system is Deimos, with dimensions of only 16 × 10 km.

The satellite which revolves around its planet closer than any other is Phobos. Phobos orbits Mars only 9377 km from the planet. This is only 2.4% the distance at which the Moon orbits the Earth.

The average atmospheric pressure on Mars, approximately 709 pascals, is equivalent to the atmospheric pressure 30 km above the surface of the Earth.

Observing data

Future dates of conjunction

1 Jul 2000
10 Aug 2002
15 Sep 2004
24 Oct 2006
6 Dec 2008
4 Sep 2011

Future dates of opposition

13 Jun 2001
28 Aug 2003
7 Nov 2005
24 Dec 2007
29 Jan 2010

Future oppositions of Mars. The distance to Earth is given in astronomical units (in parenthesis) for the date of closest approach, always within a few days of the opposition date. Note that some oppositions are more favorable. Not to scale. (Holley Bakich)

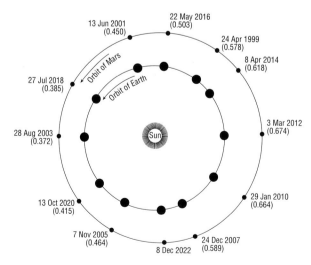

Future close conjunctions of Mars and the visible planets

	Close conjunctions	
	Date	Separation in declination
with Mercury		
	19 May 2000	67'
	10 Aug 2000	05'
	25 Jul 2002	40'
	10 Jul 2004	10'
	29 Sep 2004	51'
with Venus		
	21 Jun 2000	18'
	10 May 2002	18'
	5 Dec 2004	75'
with Jupiter		
	6 Apr 2000	66'
	3 Jul 2002	49'
	27 Sep 2004	12'
with Saturn		
	24 May 2004	95'

Recent data

At the time of writing, the most recent data concerning Mars comes to us from the NASA Mars Pathfinder mission. Launched on 4 December 1996, the spacecraft touched down on the red planet on 4 July 1997. On 5 July 1997, NASA announced that it was renaming the Mars Pathfinder lander the Carl Sagan Memorial Station, in honor of the US astronomer who passed away on 20 December 1996. On 8 August 1997, the Mars Pathfinder completed its primary mission. From the data returned, scientists have concluded that surface photographs provide strong geological and geochemical evidence that fluid water was once present on the red planet.

During the first 30 days of the mission, the imager for Mars Pathfinder returned 9669 pictures of the surface. These pictures appear to confirm that a giant flood left stones, cobbles and rocks throughout Ares Vallis, the Pathfinder landing site. In addition to finding evidence of water, the scientists confirmed that the soils are rich in iron and that suspended iron-rich dust particles permeate the Martian atmosphere.

It is known that the Martian fluvial valleys and channels are ancient features. Researchers believe that the peak of activity was about 3.5 billion years ago. Later many valleys formed. After this period, fluvial activity became localized and episodic. Cataclysmic discharges of ground water formed the huge outflow channels during this time. This water would have ponded in the northern plains of Mars. During the recent Amazonian period, only modest fluvial activity is observed. Scientists are convinced that the water that remains on Mars today is trapped, probably as permafrost and ice beneath the Martian surface.

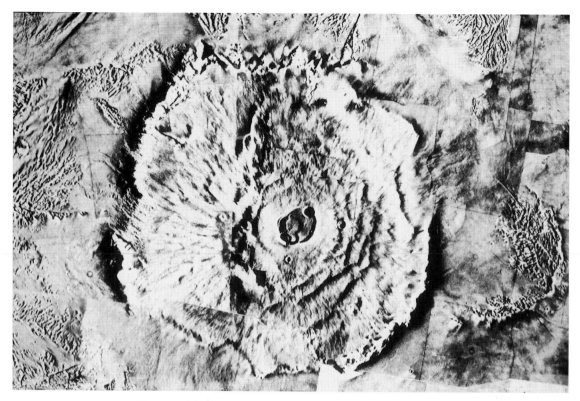

Olympus Mons, a volcano on Mars. The diameter of the base is approximately 550 km. (NASA)

Such valleys could not form today, because Mars is cold and dry. But Mars may have been significantly different in the past. At least two models of early Mars are under investigation:

(1) Mars was wet and warm. In this model, carbon dioxide released by volcanism early in Mars's history produced a greenhouse effect. Under a thick, warm atmosphere, water could flow on the surface as a liquid. An atmospheric hydrologic cycle would be possible and valley networks would form by rainfall much as they do on Earth today.

(2) Mars was already cold but wet. Even under very cold conditions, water released at springs would still have been able to flow for vast distances under an ice covering. The water would pond in low areas and freeze or infiltrate back into the subsurface. Much of the ponded and frozen flood water might be protected almost indefinitely by a covering of red soil. In this case the valley networks would not represent erosion by rainfall.

Bolstering their evidence for once-present water, the imaging team found evidence for a mineral known as maghemite – a very magnetic iron oxide. On Earth, maghemite forms in water-rich environments and could be

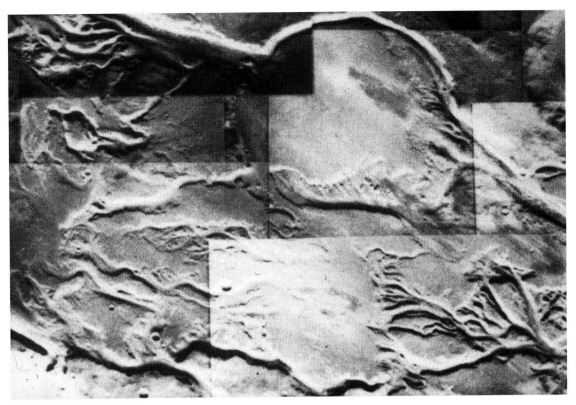

Is this evidence of early rivers on Mars? (NASA)

formed in the same way on Mars. Reddish rocks such as "Barnacle Bill," "Yogi" and "Whale rock" – pictured by the Pathfinder on the Martian surface and named by mission scientists – show evidence of extensive oxidation on their surfaces. The oxidation is possible only if water existed on the surface at some time and played an important role in the geology and geochemistry of the planet.

Of course, there is, at present, no liquid water on the surface of Mars. Several theories about the disappearing water exist, such as evaporation into space, or seepage into subsurface ice deposits or liquid aquifers, or storage at the Martian poles. At the time of writing, a new rover mission to be launched, in 2001, will attempt to determine the water's whereabouts, as well as whether the Martian environment may once have been more conducive to life.

Mars Pathfinder's camera also revealed that Mars's atmosphere is more dusty and dynamic than expected. Surprisingly, the scientists found wispy blue clouds, possibly composed of carbon dioxide, traveling through Mars's salmon-colored sky. White cirrus-like clouds, made of icy water vapor, also circulate throughout the thin Martian atmosphere. In such a thin atmosphere, these variations in the clouds of Mars were surprising.

195

Tear-drop islands,
possibly formed by
running water.
(NASA)

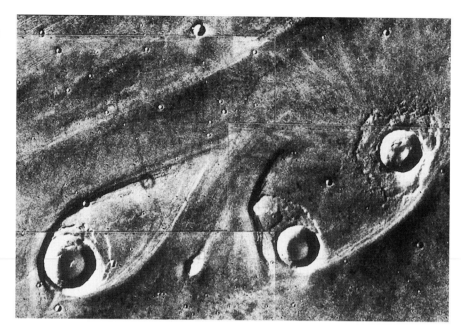

By close examination of Martian rocks like Yogi, Barnacle Bill and Scooby
Doo, the Mars Pathfinder confirmed that the rocks have been sitting on the
planet's surface for billions of years. In such a position, they endure a slow-
motion sandblasting from a usually weak, dusty Martian wind.

Satellites

	Size (km)	Mass (kg)	Density (g/cm^3)	Orbital period	Eccentricity	Inclination	Distance from planet (km)
Phobos	26×18	1.8×10^{15}	1.750	07^{h}39.2^m	0.0151	1.08°	9.377×10^3
Deimos	16×10	1.08×10^{16}	1.900	01^{d}06^{h}17.9^m	0.00033	1.79°	2.3436×10^4

**Historical
timeline**

30 July 358	Venus occulted Mars, observed from China.
14 January 375	The Chinese observed and recorded Venus occulting Mars again!
14 April 573	Mars occulted the star η Cancri; amazingly, only 36 days later – on 20 May – Venus occulted the same star.
1576–1601	Tycho Brahe recorded very accurate positions for Mars.
13 October 1590	Mars was occulted by Venus at 5:02 UT. Theoretically, this could have been seen by an observer in Earth's Southern Hemisphere, but there are no known observations of this event. Calculations were done by Belgian astronomer Jean Meeus.

1604	Johannes Kepler calculated an elliptical orbit for Mars, seriously upsetting the (presumed) harmony of the spheres.
1609	Galileo first observed Mars. The next year he wrote about observations of disk and phases (full and gibbous) indicating a spherical body illuminated by the Sun.
13 October 1659	First sketch of Mars made by Christiaan Huygens.
28 November 1659	Huygens recorded the first observation of a feature on Mars, almost certainly Syrtis Major. As he observed the feature on successive rotations, he arrived at an approximate 24-hour rotational period for Mars.
1666	Italian-born French astronomer Jean-Dominique Cassini determined the length of the Martian day as 24^h40^m.
1671	Cassini, now Director of the Paris Observatory, calculated the distance from Earth to Mars.
1672	Huygens observed a white spot at the south pole of Mars.
1698	Huygens published *Cosmotheoros*, which addresses the question of life on Mars.
1704	Cassini's nephew, Italian astronomer Giacomo Filippo Maraldi observed white spots at the poles but did not refer to them as ice caps; he noted that the south cap is not centered on the rotational pole.
1719	Maraldi raised the possibility that the white spots could be ice caps.
25 August 1719	Mars, two days from opposition, was closest to Earth. Its brightness in the sky caused panic.
1726	*Gulliver's Travels* was written by Jonathan Swift.
26–27 October 1783	Sir William Herschel observed the close passage of two faint stars near Mars; he correctly concluded that Mars has a thin atmosphere, as he could see no effect on the light of these stars when they were close to the planet.
1784	Herschel identified a 30° axial tilt for Mars; he noted the seasonal changes of the polar caps and suggested that they are composed of snow and ice.
1809	The French amateur astronomer Honoré Flaugergues, working at Viviers, noticed the presence of yellow clouds, possibly an early observation of dust clouds.
1813	Flaugergues noted variable markings on the surface of Mars, and that in the Martian spring, the polar cap shrinks rapidly; he assumed that the cap is made of layers of ice and snow and that its rapid melting proves that Mars is hotter than the Earth.

1840	The first global maps of Mars were created by Wilhelm Beer and Johann Madler; they also refined Mars's rotational period to $24^h37^m22.6^s$, within 2^s of the current value.
1858	Italian astronomer and Jesuit monk Pietro Angelo Secchi (also known as Father Secchi) sketched a map of Mars, labeling Syrtis Major the "Atlantic Canal;" Secchi believed the dark areas of Mars were seas.
1860	Emmanuel Liais proposed vegetation on Mars; he suggested that the dark areas are not seas, but rather vast tracts of vegetation.
1862	British astronomer Sir Norman Lockyer sketched Mars and agreed with Secchi that the "green" areas of Mars are oceanic.
1863	The first color sketches of Mars were drawn by Father Secchi.
1867	First attempts were made to detect oxygen and water vapor spectroscopically, by French astronomer Pierre Jules Janssen and British astronomer Sir William Huggins; results were inconclusive.
1868	British astronomy popularizer Richard Anthony Proctor published *The Lands and Seas of Mars, from 27 Drawings by Mr Dawes*. This was a record of observations of Mars conducted by William Rutter Dawes from 1852 to 1865. Proctor's choice of the zero meridian of Mars survives.
1869	Father Secchi referred to "canali," Italian for channels.
1873	French astronomer Camille Flammarion ascribed the red color of Mars to vegetation.
1877	Reports of "canali" were made by Giovanni Virginio Schiaparelli; Schiaparelli developed a nomenclature for the features observed on Mars.
11 August 1877	US astronomer Asaph Hall discovered Deimos.
17 August 1877	Asaph Hall discovered Phobos.
1879	Schiaparelli reported an observation of double canali; this came to be known as "gemination."
1894	Percival Lowell built an observatory in the territory of Arizona, at Flagstaff; he made his first observations of Mars, which was the main reason for the construction of the observatory.
1894	Edward Emerson Barnard reported on his observations of Mars including his complete failure to detect canals.
1895	The first edition of Percival Lowell's *Mars*, his first book on the red planet, was published.

Mars, showing the phenomenon of gemination. Schiaparelli, Giovanni Virginio, *Il pianeta marte*, Milan, 1893. (Linda Hall Library)

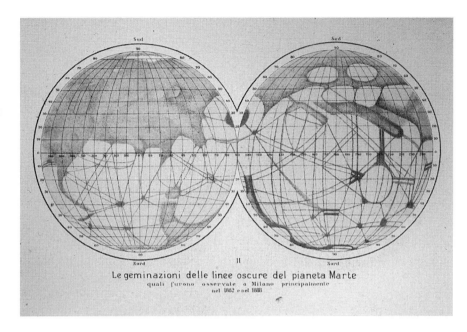

Le geminazioni delle linee oscure del pianeta Marte
quali furono osservate a Milano principalmente
nel 1882 e nel 1888

1898	*War of the Worlds* by Herbert George Wells published; the book was serialized during the previous year in England and the United States.
16 January 1901	US astronomer William Henry Pickering reported a shaft of light seen to project from Mars.
December 1906	Percival Lowell's second book on Mars, *Mars and its Canals* published.
December 1908	Percival Lowell's third book on Mars, *Mars as the Abode of Life* published; this was reprinted from a series of articles in *Century Magazine* from November 1907 to June 1908.
1909	Percival Lowell's fourth and final book on Mars, *The Evolution of Worlds*, published.
1909	US astronomer George Ellery Hale, using the Mt Wilson 152-cm reflector, saw ". . . not a trace" of canals.
1911	A meteorite killed a dog in Nakhla, Egypt. In 1986, tests on this meteorite determined that it originated on Mars.
1912	Edgar Rice Burroughs published *A Princess of Mars*, the first of eleven "John Carter on Mars" novels; he uses Schiaparelli's nomenclature; Burroughs's Martians have green skin.
1925	US astronomer Donald H. Menzel, studying photographs of Mars taken at different wavelengths, concluded that the air pressure on Mars is less than 6687 pascals.

199

1926	US astronomer Walter Sydney Adams determined spectroscopically that Mars is "ultra-arid."
1927	Large temperature differences between the day and night sides of Mars were measured by William Weber Coblentz and Carl Otto Lampland; this is taken to be a sign of a very thin atmosphere.
30 October 1938	"War of the Worlds" broadcast by Orson Welles (*Grovers Mill, New Jersey landings*). Estimates suggest that 1.2 million of the 6 million listeners thought it was real.
1947	Using infrared spectroscopy, Dutch-born US astronomer Gerard Peter Kuiper detected carbon dioxide on Mars, but no oxygen.
1950	US science fiction writer Ray Bradbury published *The Martian Chronicles*.
1 November 1962	USSR probe Mars 1 launched, but radio contact lost after 95 500 000 km; passed within 193 000 km of Mars on 19 June 1963; now orbiting the Sun.
5 November 1964	US probe Mariner 3 lost when its protective shroud failed to eject as the craft was placed into interplanetary space.
28 November 1964	US probe Mariner 4 launched; flew to within 9846 km of Mars on 14 July 1965; took 22 photographs covering approximately 1% of the Martian surface.
30 November 1964	USSR probe Zond 2 launched; passed within 1609 km of Mars in 1965, but sent no data.
18 July 1965	USSR probe Zond 3 launched; sent back 25 pictures of the far side of the Moon as a communications test; no data from Mars returned.
24 February 1969	US probe Mariner 6 launched; flew by Mars on 31 July 1969, passing within 3437 km of the planet's equatorial region; sent back 200 photographs; measured surface and atmospheric temperature, surface molecular composition and atmospheric pressure; now orbiting the Sun.
27 March 1969	US probe Mariner 7 launched; flew by Mars on 5 August 1969, passing within 3551 km of the planet's south polar region; sent back 200 photographs; similar experiments to Mariner 6; now in solar orbit.
10 May 1971	USSR probe Mars 2 launched; entered Mars orbit 27 November 1971; studied surface and atmosphere; landing capsule touched down 27 November, but no data returned; orbiter transmitted data until March 1972.
28 May 1971	USSR probe Mars 3 launched; identical to Mars 2; entered Mars orbit 2 December 1971; lander touched

	down 2 December, transmitted data for 2 minutes; orbiter transmitted data until March 1972.
30 May 1971	US probe Mariner 9 launched; arrived at Mars on 13 November 1971; mapped entire surface and gathered data on atmospheric composition, density and temperature; sent back 7329 photographs; after depleting its supply of control propellant, the spacecraft was turned off 27 October 1972.
13 November 1971	Mariner 9 became the first artificial satellite of another planet.
21 July 1973	USSR probe Mars 4 launched; overshot Mars, flew within 2250 km on 10 February 1974; transmitted some photographs of Mars as it flew by; now orbiting the Sun.
25 July 1973	USSR probe Mars 5 launched; identical to Mars 4; entered Mars orbit 2 February 1974; transmitted data for 3 days.
5 August 1973	USSR probe Mars 6 launched; flew past Mars 12 March 1974; lander deployed as probe flew by; lander touched down 12 March, but stopped sending data 02^m28^s into parachute descent; Mars 6 now orbiting the Sun.
9 August 1973	USSR probe Mars 7 launched; identical to Mars 6; spacecraft flew past Mars 9 March 1974 and deployed lander; lander missed Mars by 1287 km.
20 August 1975	US space probe Viking 2 was launched toward Mars (prior to the launch of Viking 1); arrived at Mars 7 August 1976; landed on Mars 3 September 1976; transmitted data for 3.5 years.
9 September 1975	US space probe Viking 1 was launched toward Mars; arrived at Mars 19 June 1976; landed on Mars 20 July 1976; transmitted data for 6.5 years.
8 April 1976	Mars occulted the 2.98 magnitude star ϵ Geminorum (Mebsuta). The author, then a graduate student at Michigan State University, won $50 for the best observing project of the year.
11 May 1984	The most recent transit of Earth, as seen from Mars, occurred.
12 July 1988	USSR probe Phobos 2 launched; arrived at Mars 29 January 1989; identified water vapor in atmosphere; took photographs of Mars and Phobos; stopped transmitting on 27 March 1989; unable to deploy hopping lander on Phobos, April 1989.
17 July 1988	USSR probe Phobos 1 launched; traveled 11% of distance to Mars until accidentally turned off on 2 September 1988; now in orbit around the Sun.
4 December 1996	US probe Mars Pathfinder launched aboard a Delta 2 rocket.

| 4 July 1997 | US probe Mars Pathfinder arrived on the surface of Mars. Pathfinder deployed a small rover called Sojourner to explore the Martian landscape. |
| 10 November 2084 | The next transit of Earth, as seen on Mars, will occur. |

Jupiter

Physical data

Size   142 984 km*

Mass 1.900×10^{27} kg

Escape velocity   59.366 km/s (213 718 km/hr)

Temperature range   Minimum   Average   Maximum
$-163\,°C$ $-121\,°C$ increases with depth

Oblateness   0.06481

Surface gravity   24.51 m/s² (equatorial)
26.36 m/s² (polar)†

Volume 1.377×10^{15} km³
1266 times that of Earth‡

Magnetic field strength and orientation   Reaching a distance of more than 1 600 000 km from the planet, the magnetosphere of Jupiter is the most extensive of any of the planets. The strong field is generated by Jupiter's rapid rotation coupled with an interior of metallic liquid hydrogen, which acts in a way similar to the liquid iron core of the Earth.

Average surface field (tesla)	0.0004
Dipole moment (weber-meters)	3.14×10^{20}
Tilt from planetary axis	10°
Offset from planetary axis (planet radii)	0.1

Albedo (visual geometric albedo)   0.52

Density (water = 1)   1.33 g/cm³

Solar irradiance   51 watts/m²

Atmospheric pressure   varies with depth, >10 132 500 pascals

Composition of atmosphere		
	Molecular hydrogen (H_2)	>81%
	Helium (He)	>17%
	Methane (CH_4)	0.1%
	Water vapor (H_2O)	0.1%
	Ammonia (NH_3)	0.02%
	Ethane (C_2H_6)	0.0002%
	Phosphine (PH_3)	0.0001%
	Hydrogen sulfide (H_2S)	<0.0001%
	Acetylene (C_2H_2)	0.000 003%
	(continued overleaf)	

* Polar diameter 133 717 km.
† The rapid rotation of Jupiter causes the planet to be flattened.
‡ If Jupiter were spherical, it would hold 1408 Earths.

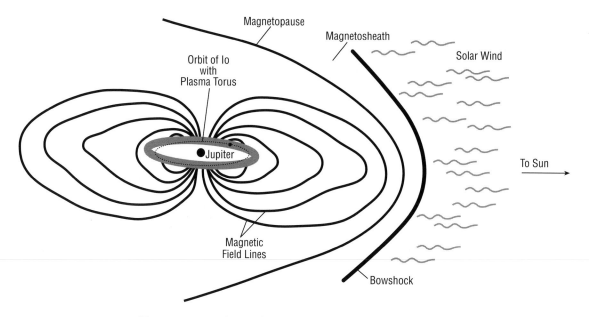

The magnetosphere of Jupiter. (Holley Bakich)

Composition of atmosphere (cont.)	Monodeuteromethane (CH_3D)	0.000002%
	Hydrogen cyanide (HCN)	0.0000001%
	Ethylene (C_2H_4)	0.0000001%
	Hydrazine (N_2H_4)	<0.0000001%
	Methylamine (CH_3NH_2)	<0.0000001%
	Germane (GeH_4)	0.00000006%
	Carbon monoxide (CO)	0.00000001%

Maximum wind speeds   531 km/hr

Outstanding cloud features

The most visible feature within the clouds of Jupiter is the anticyclonic high-pressure region named the Great Red Spot. Possibly observed since 1664, and definitely since the nineteenth century, the Great Red Spot may not be a permanent feature in Jupiter's atmosphere. Indeed, during the twentieth century, the Great Red Spot has steadily decreased in size. In addition to the Great Red Spot, other spots of various colors and sizes may often be seen.

The overall appearance of Jupiter is that of a planet whose atmosphere is divided into bands of various colors which are oriented parallel to the planet's equator. These bands, of course, represent clouds. The lighter colored bands are known as "zones" and the darker colored ones are called "belts." Within the zones and belts are eddies which may produce temporary spots or streaks within or between Jupiter's cloud bands.

To standardize the nomenclature related to Jupiter's cloud bands, the scheme shown in the diagram is often used:

Jupiter. Cassini, Jean-Dominique, "Relation du retour d'une grande tache permanente dans la planète de Jupiter …" *Journal des Scavans*, **3**: 43–50, Amsterdam, 1678. (Linda Hall Library)

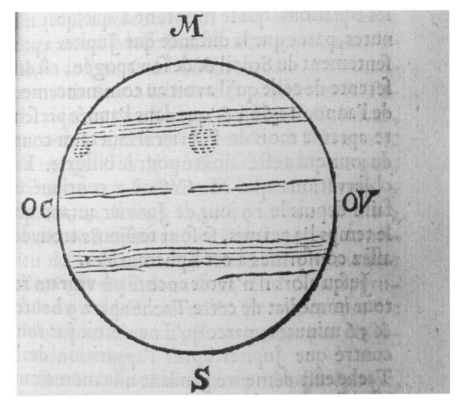

Typical observing nomenclature for Jupiter. (Holley Bakich)

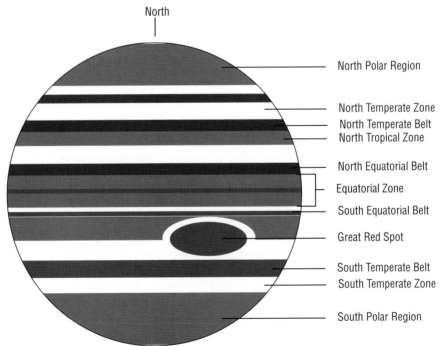

North

North Polar Region

North Temperate Zone
North Temperate Belt
North Tropical Zone

North Equatorial Belt

Equatorial Zone

South Equatorial Belt

Great Red Spot

South Temperate Belt
South Temperate Zone

South Polar Region

Orbital data

Period of rotation 0.413 54 days
9.9250 hours
$0^d9^h55.5^m$

Period of revolution (sidereal orbital period) 11.8626 years
4332.71 days
$11^y315^d1.1^h$

Synodic period $398^d21^h07.2^m$

Equatorial velocity of rotation 12572 m/s (45259.5 km/hr)

Velocity of revolution 13.07 km/s (47052 km/hr)

Distance from Sun Average 5.2028 AU
778330000 km
Maximum 5.4570 AU
816355600 km
Minimum 4.9500 AU
740509500 km

Distance from Earth Maximum 6.47 AU
968460580 km
Minimum 3.93 AU
588404520 km

Apparent size of Sun (average) 0.102°

Apparent brightness of Sun $m_{vis} = -23.1$

Orbital eccentricity 0.0483

Orbital inclination 1.308°

Inclination of equator to orbit 3.13°

Observational data

Maximum angular distance from Sun 180°

Brilliancy at opposition Maximum −2.9
Minimum −2.0

Angular size* Maximum 50.11″
Minimum 30.467″

Early ideas

The Sumerians recognized the planet Jupiter by the names Sag-nae-gar and Mul-babbar. The Babylonians called it Nibiru-Marduk and Udaltar.

The Babylonians noticed a periodicity with regard to the motions of the planets. For example, they calculated that 71 years equaled 65.01 synodic periods

---

* This measurement is the apparent angular equatorial diameter of Jupiter, measured in seconds of arc, as seen from Earth. Note: Jupiter's polar diameter is 93% of its equatorial diameter.

Jupiter, showing belts. Riccioli, Giovanni Battista, *Almagestum novum*, Bologna, 1651. (Linda Hall Library)

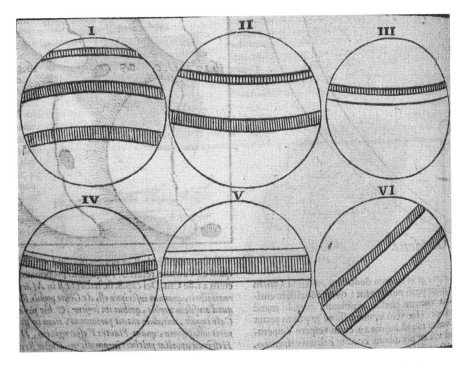

of Jupiter (essentially the round number 65) and also 5.99 sidereal periods (essentially 6). Thus, to a surprisingly high degree of accuracy, the Babylonians could foretell the longitudinal position of Jupiter on certain dates by looking at the data earlier observers had accumulated 71 years in the past. They also found that they could predict positions in a similar way using an 83-year period. Such a compilation has come to be known as a "goal-year-text." The Babylonians began this detailed record keeping earlier than 250 BC. Tablets show that observational records were still being produced in the late period, up to 50 BC. Some tablets divided the ecliptic into sections. One Jupiter tablet divided the ecliptic into four areas: one fast, one slow, two medium. This denoted the speed with which Jupiter was observed to move against the starry background.

The Chinese called the Sun, Moon and five visible planets the "Seven Luminaries." Observations of the seven were for astrological purposes. Conjunctions of any two luminaries foretold a variety of events. Most, it must be stated, were bad: "If Mars is sickle-shaped and retrogresses, there will be a military defeat."

The Chinese used their observations of Jupiter, which they called Mu xing, in an additional way. They sectioned the sky into 12 "Jupiter stations." They observed that the sidereal period of Jupiter is roughly 12 years. That is, they noticed that Jupiter took approximately 12 years to move completely through the zodiac, where it would begin the cycle anew. This value was in error by about 50 days, or slightly more than one percent – more than accurate enough for their purposes. Chinese astrologers associated each of these with one character from a sequence of twelve animals used in their sixty-name cycle. Each name in this cycle corresponded to some part of the country.

The Jovian system.
Doppelmayr,
Johann Gabriel,
Atlas Coelestis,
Nuremberg, 1742.
(Linda Hall Library)

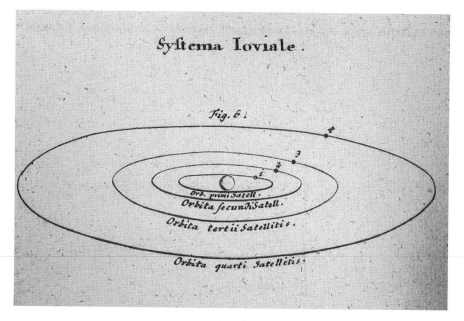

An interesting sidelight to astronomical history has come about from the study of ancient Chinese observations. Apparently, the Chinese astronomer Gan De observed the satellite Ganymede in the summer of 364 BC, 1974 years before Galileo saw it through his simple telescope. This was an incredible observation, requiring not only superb eyesight, but also a night of the most excellent observing conditions.

In 1610, the Italian astronomer Galileo Galilei discovered the four large moons of Jupiter. This important find will be discussed in a later section. In 1675, the Dutch astronomer Ole Romer made a number of observations of the Galilean satellites as they disappeared into the shadow of Jupiter. It had been noticed that the predicted timings of these events differed from observation based upon where Jupiter was in its orbit. Romer correctly deduced that light had a finite velocity and the travel time of the light from these eclipses was causing the errors. His value for the speed of light was 75% of that which is currently accepted – not a bad first approximation.

In the latter half of the nineteenth century, after spectroscopy revealed the processes occurring within stars and later, when we learned something about star formation, the question was raised as to whether Jupiter was a failed star. This was based upon the observation that Jupiter has essentially the same chemical makeup as the Sun. As more was learned, however, astronomers realized that the mass of Jupiter was far too low to allow nuclear fusion at its core. The least massive star which can support nuclear reactions has a mass approximately 0.08 of the Sun. Jupiter's mass is approximately 0.001 that of the Sun. Therefore, one would have to crush 80 Jupiters together to make even the least massive viable star.

On 9 September 1892, US astronomer Edward Emerson Barnard (1857–1923) discovered the fifth satellite of Jupiter, now known as Amalthea.

Jupiter, 24 October 1870, contrasting a drawing (top) with a photograph by Warren de la Rue. Browning, John, "On a photograph of Jupiter" *Monthly Notices of the Royal Astronomical Society*, **31** (no. 2): 33–4, London, 1870. (Linda Hall Library)

Comet
Shoemaker–Levy 9
confirmation
image. (James V.
Scotti)

He was observing with what was then the largest telescope in the world, the 91-cm refractor at the Lick Observatory. Even so, it was a fantastic discovery considering the brilliance of Jupiter and the faintness of the satellite. Barnard has the distinction of making the last discovery of a satellite with the naked eye.

Important concepts

Comet Shoemaker–Levy 9

Comet Shoemaker–Levy 9 was discovered in March 1993, by Eugene and Carolyn Shoemaker and David Levy. In May 1993, it was recognized for the first time that Shoemaker–Levy 9 was on a collision-course with Jupiter, the impact to take place in mid-July 1994.

Since this was the first observation of Comet Shoemaker–Levy 9, its earlier orbit was very uncertain. When astronomers extrapolated backward, it seemed that Comet Shoemaker–Levy 9 had orbited Jupiter for many years. A leading train of thought points to the possibility that the comet was captured between 1920 and 1930.

On 7 July 1992, Comet Shoemaker–Levy 9 passed within 1.286 Jupiter radii (approximately 91 939 km). At that time, the comet was fragmented into at least 20 individual pieces. Since several of the impacts on Jupiter were multiple, further break-ups as the comet neared the giant planet have been theorized. Soon after fragmentation a great deal of dust was observed. The amount of dust diminished over time. No gas was ever observed within the coma of any of the cometary pieces. At the distance of Jupiter, this is not

210

unusual. The diameter of the original comet nucleus has been estimated at between 4 and 5 km. From calculations based upon the energy released during each impact, the minimum fragment diameter has been estimated at 350 m. The largest fragment measured 1–2 km in diameter.

The impact sites were not directly observable from the Earth. Direct views were available from the Galileo spacecraft, which was 240 000 000 km from Jupiter, and also from Voyager 2, then a lofty 6 200 000 000 km from Jupiter. From Earth, indirect observations of the impacts were attempted by monitoring the light intensity from Jovian moons in order to detect possible reflections of the light emitted by the descending bolide, the subsequent explosion and the beginning ascent of the resulting fireball.

The Galileo spacecraft observed the K, N, V and W impacts with an image resolution of 2.5 seconds of arc. As an example, the K fragment explosion showed a 5 s rise in brightness to a level equal to 15% of Jupiter's total light output. It remained at this level for 49 s. The explosion of the W fragment, which was observed by the Hubble Space Telescope, showed a total luminous energy output of $2.1–4.4 \times 10^{26}$ ergs, with a fireball temperature of 18 000 K. This allowed scientists to estimate a minimum impactor radius of 250–350 m. It is believed that the fragment must have been significantly larger, since much of the kinetic energy would not transform into luminous energy, but would be deposited inside Jupiter's atmosphere.

Based upon detailed study of all impacts, the most plausible scenario begins with the fragments of Comet Shoemaker–Levy 9 striking Jupiter, with their impacts directed along a 45° tunnel. The lower portion of each fragment continued to move downward even as the upper part of the fireball began to rise. Most of the kinetic energy was deposited deep in Jupiter's atmosphere. It is thought that the biggest fragments reached the H_2O cloud layer.

The fireballs provided some fascinating data. From spectroscopic studies, we know that most of the material in the fireballs was from the atmosphere of Jupiter. The initial fireball temperature was above 10 000 K, but the cooling was rapid: 400–700 K after 3–10 min. Shortly after impact, material was moving in ballistic orbits and reached an altitude of about 3200 km. The maximum height of each fireball was reached after about 500 s. Then the material in each fireball rained down on the upper atmosphere causing heating.

After about 30 min the fireballs had been transformed into black clouds (also referred to as dark spots or plumes), easily visible at all wavelengths and observable with even very small telescopes. These structures resembled "pancakes," and formed in the stratosphere of Jupiter where the pressure was approximately 100 pascals. Each of the plumes had a diameter greater than 10 000 km. The major plume structures were still clearly visible in late September.

The Hubble Space Telescope Faint Object Spectrograph detected many gaseous absorptions associated with the impact sites of Comet Shoemaker–Levy 9 and followed their evolution during the month following the event. Most surprising were the strong signatures from sulfur-bearing compounds like diatomic sulfur (S_2), carbon disulfide (CS_2) and hydrogen sulfide (H_2S). Ammonia (NH_3) absorption was also detected. The S_2 absorptions seemed to fade after a few days, while the NH_3 absorptions at first got stronger, and then started fading after about one month.

The high-speed easterly and westerly jets turned the dark plumes at the impact sites into striking curlicue features. Although individual impact sites were still visible a month later despite the shearing effects of the atmosphere, the fading of Jupiter's scars has been substantial and it now appears that Jupiter will not suffer any permanent changes from the explosions.

Hubble's ultraviolet observations showed the motion of very fine impact debris particles suspended high in Jupiter's atmosphere. This provided the first information ever obtained about Jupiter's high-altitude wind patterns. At lower altitudes, the impact debris follows east–west winds driven by sunlight and Jupiter's own internal heat. By contrast, winds in the high Jovian stratosphere move primarily from the poles toward the equator because they are driven mainly by auroral heating from high-energy particles.

Hubble detected unusual auroral activity in Jupiter's northern hemisphere just after the impact of the K fragment. This impact completely disrupted the radiation belts which have been stable over the last 20 years of radio observations. Aurorae are common on Jupiter because energetic charged particles needed to excite the gases are always trapped in Jupiter's magnetosphere. However, this new feature seen by Hubble was unusual because it was as bright or brighter than the normal aurora, short-lived and outside the area where Jovian aurorae are normally found. Astronomers believe the K impact created an electromagnetic disturbance that traveled along magnetic field lines into the radiation belts. This scattered charged particles, which normally exist in the radiation belts, into Jupiter's upper atmosphere.

Recently, A. L. Sprague of the Lunar and Planetary Laboratory, Tucson, AZ, and his colleagues reported that spectroscopic measurements in the impact plumes of the R and W fragments of Comet Shoemaker–Levy 9, made from the Kuiper Airborne Observatory, revealed two H_2O emission features. The team's calculations show that the R impactor had a diameter of approximately 300 m, containing H_2O ice equivalent to a sphere 60 m in diameter.

Research into this event persists, data continues to be analyzed and theories abound. But these tell only part of the story. This author remembers the *feeling* during the summer of 1994. I remember the excitement of the researchers and the absolute awe of the general public. I encourage you to take a moment and think back on what was, in my estimation, the astronomical event of the century.

The table on the following page provides the date and time of the impact of each of the fragments of Comet Shoemaker–Levy 9.

Fragment	Date (July 1994)	UT time (h:m:s)
A	16	20:11:00
B	17	02:53:00
C	17	07:12:00
D	17	11:54:00
E	17	15:11:00
F	18	00:33:00
G	18	07:33:32
H	18	19:31:59
K	19	10:24:14
L	19	22:16:48
N	20	10:29:17
P2	20	15:23:00
Q2	20	19:44:00
Q1	20	20:13:00
R	21	05:34:00
S	21	15:15:00
T	21	18:10:00
U	21	21:55:00
V	22	04:23:00
W	22	08:06:12

The Great Red Spot

For more than a century, humans have been observing the Great Red Spot. In fact, it may have been observed in 1664, by the English astronomer Robert Hooke. The Great Red Spot is an anticyclonic (high-pressure) storm located 22° south of Jupiter's equator. The closest analogy would be to a terrestrial hurricane. Since it is anticyclonic in Jupiter's southern hemisphere, the rotation of the Great Red Spot is counterclockwise, with a period of about 6 days. For comparison, a hurricane in Earth's southern hemisphere rotates clockwise because it is a low-pressure system. The Spot itself is enormous: it has a north–south width of 14 000 km and a variable east–west width of 24 000–40 000 km. This means that three Earths could fit within the boundaries of the Great Red Spot. The clouds associated with the Great Red Spot appear to be about 8 km above neighboring cloud tops.

The Coriolis effects that are responsible for cyclones and anticyclones on Earth are greatly magnified on Jupiter. This is understandable when we compare the Earth's rotational period of approximately 24 hours with Jupiter's rotation, once every 10 hours, approximately. This difference alone does not account for the persistence and size of the Great Red Spot. There are other features similar to the Great Red Spot on the surface but none are as large.

Presumably the persistence of the Great Red Spot is related to the fact that it never comes over land, as in the case of a hurricane on Earth. Thus, on

The Great Red
Spot. The spot
measures
approximately
30 000 km in length.
(NASA)

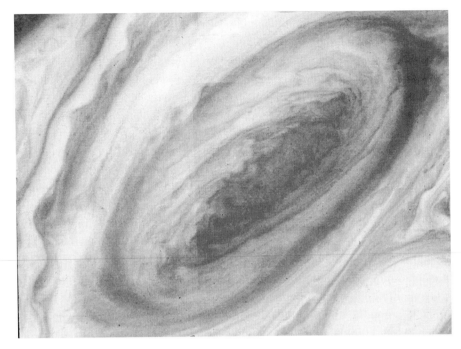

Jupiter, the friction which would dissipate such a structure is missing. Another factor adding to the persistence of the Red Spot is that its motion is driven by Jupiter's internal heat source. Computer simulations suggest that such large disturbances may be stable on Jupiter, and that stronger disturbances tend to absorb weaker ones, which may explain the size of the Great Red Spot.

One unanswered question about the Great Red Spot is "Why is the Spot red?" It has been suggested that certain compounds of phosphorus are responsible for the reddish-brown hue, but this remains somewhat speculative.

Finally, it is not known how long the Great Red Spot will last. Significant changes in size have taken place during the twentieth century. At the time of writing, the Spot is approximately half as large as it was 100 years ago.

Galileo and his satellites

The Italian astronomer Galileo Galilei made a number of discoveries with his early telescopes. He saw irregularities on the surface of the Moon and spots on the surface of the Sun. He noted that Venus went through a cycle of phases, similar to the Moon. He saw thousands of stars which had never been seen by human eye. But the most miraculous result of his exploration of the heavens was made on 7 January 1610. On that night, Galileo looked at Jupiter through his telescope and saw three little stars in a straight line, two on one side of Jupiter and one on the other. The next night, the stars were still there,

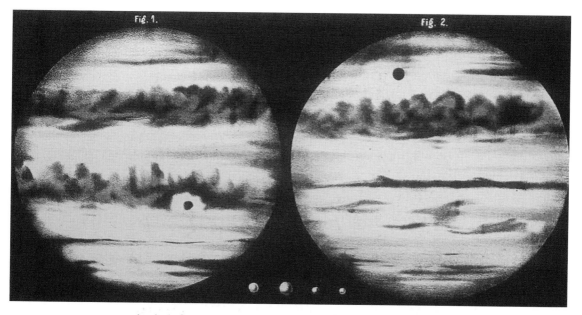

Jupiter. Secchi, Angelo, "Jupiter" *Sirius*, **9** (no. 6), opp. p. 148, Leipzig, 1876.
(Linda Hall Library)

but their positions were different. This continued until 13 January 1610, when
Galileo noticed a fourth star.

After watching them for a number of weeks, Galileo came to the conclusion
– and he could scarcely believe it – that the four "stars" were actually satellites revolving around Jupiter in the same way that the Moon circles the Earth.
These were the first solar system objects discovered that were invisible to the
unaided eye. It was a tremendous discovery, and one which helped to estab-
lish Copernicus's ideas as the true model of the solar system.

It had been 67 years since Copernicus's *De Revolutionibus Orbium Coelestium*
had been published, but conservative thought still held that the Earth, not the
Sun was the center of the solar system. Galileo's telescope proved that there
were bodies which did not revolve around the Earth. They revolved around
Jupiter. The astonished world learned of Galileo's discovery in March 1610,
when he published *Sidereus Nuncius*.

This book was dedicated to Galileo's patron Cosimo II de Medici, Grand
Duke of Tuscany. Galileo also desired to name the new planets after the Grand
Duke, but what to call them? He wrote to ask if he should call them the Cosmic
stars (from Cosimo) or the Medicean stars (from Medici). The Grand Duke
answered that he preferred the second designation. Galileo sent him a copy of
his book and the actual telescope with which he had made the discoveries.

It is difficult, nearly 400 years after the fact, to convey the furor that this
discovery caused. Kings and princes of Europe requested telescopes, letters
were sent to Galileo requesting that when new stars were discovered, they
be named after this monarch or that patron. Galileo was hailed as the king of
scientists. But not all the tumult was positive.

Galileo Galilei.
(Michael E. Bakich
collection)

Supporters of Aristotle's model of the universe and many religious con-
servatives within the Church opposed Galileo. Some claimed that Galileo had
invented the objects or that it was a hallucination or a joke. Many of Galileo's
detractors obstinately refused to look through a telescope at all. One contemporary, named Cremonino, claimed that to do so would be to commit
"treason against Aristotle." And on it went, causing no end of trouble for
Galileo. The story is a sad but famous one.

There were also scientific opponents to Galileo. Simon Marius claimed

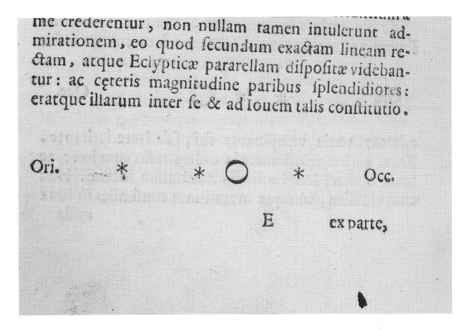

me crederentur, non nullam tamen intulerunt admirationem, eo quod fecundum exactam lineam rectam, atque Ecliypticæ pararellam difpofitæ videbantur: ac ceteris magnitudine paribus fplendidiores: eratque illarum inter fe & ad Iouem talis conftitutio.

Ori.   ∗   ∗ ◯ ∗   Occ.

E   ex parte,

precedence in the discovery of Jupiter's "companions." It was Marius who dubbed them "satellites," a Latin word for individuals who cluster about some rich and powerful person in hope of being invited to dinner or of receiving gifts. Marius saw a parallel in the way the new worlds seemed to hover around Jupiter. He also gave each a proper name, signifying a connection to the mythical figure of Jupiter, king of the Roman gods. Io, Europa and Callisto were named after nymphs with whom Jupiter fell in love at one time or another. Ganymede was named for a handsome young man whom Jupiter took to Mt Olympus to serve as cupbearer to the gods.

The rings of Jupiter

The ring system of Jupiter is quite different from that of Saturn. Unlike the Saturnian system, Jupiter's rings are formed by charged particles of various sizes. Most of these particles are very tiny (about 1 micron across). There are two forces that are exerted on these particles by Jupiter, a gravitational force and an electromagnetic force. The gravitational force is stronger than the electromagnetic force for particles larger than 1 micron and this force provides the centripetal acceleration that is required to keep these particles in orbit around Jupiter.

Throughout their lifetime these particles are ground down by interactions with the energetic particles that are abundant in Jupiter's magnetosphere. Over time, they become so small (about 0.03 micron) that the electromagnetic force overpowers the gravitational force and the particles leave the rings and fall into Jupiter's atmosphere. The lifetime of these particles is about 1000 years. However, Jupiter's rings are a permanent feature because these tiny

A page from Galileo's journal. The middle image is his first observation of all four "Galilean" satellites. Galilei, Galileo, *Sidereus Nuncius*, Venice, 1610. (Linda Hall Library)

OBSERVAT. SIDEREAE

Ori.   ✳ ✳○ ✳   Occ.

Stella occidentaliori maior, ambæ tamen valdè conspicuæ, ac splendidæ : vtra quæ diftabat à Ioue scrupulis primis duobus; tertia quoque Stellula apparere cępit hora tertia priùs minimè conspecta, quæ ex parte orientali Iouem ferè tangebat, eratque admodum exigua. Omnes fuerunt in eadem recta, & secundum Eclypticæ longitudinem coordinatæ.

Die decimatertia primum à me quatuor conspectæ fuerunt Stellulæ in hac ad Iouem conftitutione. Erant tres occidentales, & vna orientalis; lineam proximè

Ori.   ✳ ○✳✳✳   Occ.

rectam conftituebant ; media enim occidétalium paululum à recta Septentrionem verfus deflectebat. Aberat orientalior à Ioue minuta duo : reliquarum, & Iouis intercapedines erant singulæ vnius tantum minuti. Stellæ omnes eandem præ se ferebant magnitudinem ; ac licet exiguam, lucidiffimæ tamen erant, ac fixis eiufdem magnitudinis longe fplendidiores.

Die decimaquarta nubilofa fuit tempeftas.

Die decimaquinta, hora noctis tertia in proximè depicta fuerunt habitudine quatuor Stellæ ad Iouem ;

Ori.   ○ ✳ ✳ ✳   ✳   Occ.

occidentales omnes: ac in eadem proxim recta linea difpofitæ; quæ enim tertia à Ioue numerabatur paululum

particles are regenerated continually. Jupiter's intricate, swirling ring system is formed by dust kicked up as interplanetary meteoroids smash into the giant planet's four small inner moons, according to data from NASA's Galileo spacecraft.

In these impacts, the meteoroid is going so fast it buries itself deep in the moon, then vaporizes and explodes, causing debris to be thrown off at such high velocity that it escapes the satellite's gravitational field. If the moon is too big, dust particles will not have enough velocity to escape the moon's

Title page of Galileo's *Sidereus Nuncius,* Venice, 1610. (Linda Hall Library)

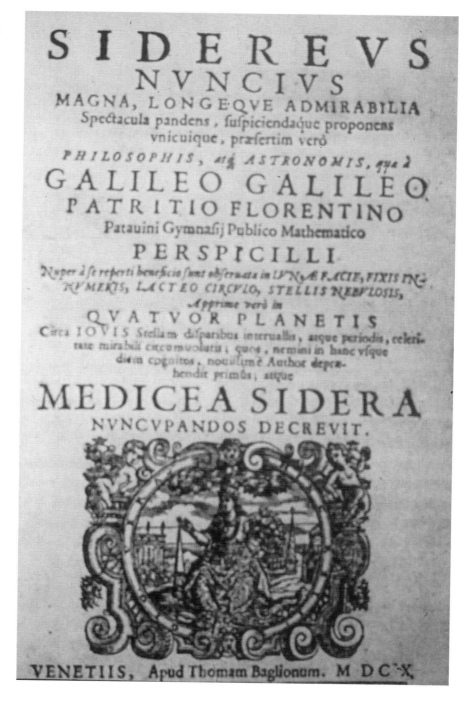

Jupiter and the Crescent Moon, nearing occultation. (Photograph by Vic Winter)

gravitational field. With a radius of less than 15 km and an orbit that lies at the periphery of the main ring, tiny Adrastea is most perfectly suited for the job.

In the late 1970s NASA's two Voyager spacecraft first revealed the structure of Jupiter's rings: a flattened main ring and an inner, cloud-like ring, called the halo, both composed of small, dark particles. One Voyager image seemed to indicate a third, faint outer ring.

Data from the Galileo spacecraft revealed that this third ring, known as the gossamer ring because of its transparency, consists of two rings. One is embedded within the other, and both are composed of microscopic debris from two small moons, Amalthea and Thebe. NASA scientists further believe that the main ring is composed of material from the satellites Adrastea and Metis. Galileo took three dozen images of the rings and small moons during three orbits of Jupiter in 1996 and 1997. These images showed that the rings contain very tiny particles resembling dark, reddish soot. Unlike Saturn's rings, there are no signs of ice in Jupiter's rings.

Interesting facts

From Jupiter, the Sun appears 3.6% as bright as from Earth.

The four large satellites of Jupiter were discovered in 1610. It was to be 282 years before the fifth satellite (Amalthea) was discovered, by Edward Emerson Barnard, on 9 September 1892.

The Galilean satellites were named by the German-born Dutch astronomer Simon Marius, who also claimed to have discovered them.

Barnard discovered Amalthea. It was his right to name it, but Barnard could not find a name that he liked prior to his death. Soon after Barnard's discovery, the name Amalthea had been suggested by the French astronomer Camille Flammarion. Because of Barnard's indecision, "Amalthea" was the name that came into general use by the astronomical community.

Amalthea was the last satellite discovered by visual means. Barnard found it visually using the 91-cm refracting telescope of the Lick Observatory in California.

Jupiter is the second most reflective planet, with a geometric albedo of 0.52.

The largest satellite in the solar system is Ganymede, with a diameter of 5268 km.

Jupiter's polar diameter is 93% that of its equatorial diameter.

The Galileo spacecraft carried a 20-watt transmitter. From Earth, the radio signals received from Galileo were approximately one billion times weaker than a transistor radio heard from a distance of 5000 km.

Jupiter is more than 317 times as massive as the Earth.

Sinope has the largest orbit of any known satellite. It revolves around Jupiter at a distance of 23 700 000 km. This is more than 61 times the distance at which the Moon orbits the Earth.

At Jupiter's equator, the speed of rotation is 45 259.5 km/hr. This is more than 27 times as fast as the Earth at its equator.

Ganymede is the most massive satellite in the solar system, possessing nearly 2.5% the mass of the Earth.

Jupiter is the largest planet. Not only would 1266 Earths fit inside, but the eight other planets would only comprise 68.8% of Jupiter's volume.

The eight outer moons of Jupiter are named in an interesting way. Those satellites with direct orbital motion have names ending in "a" (Elara, Himalia, Leda, Lysithea). Those with retrograde orbits have names ending in "e" (Ananke, Carme, Pasiphae, Sinope).

The largest single structure in the solar system is the magnetosphere of Jupiter. If it were visible from Earth, it would appear larger than the Full Moon.

Relative sizes of the
planets, to scale.
(Holley Bakich)

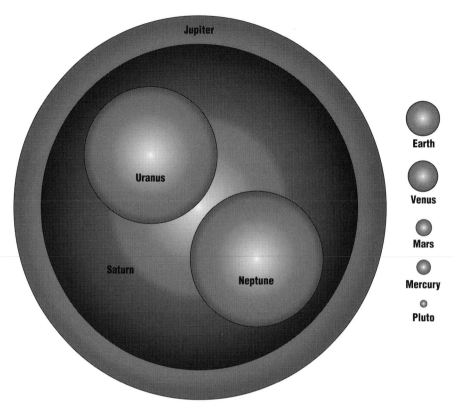

The faintest of the satellites of Jupiter is Leda. At opposition, it has a mean visual magnitude of 20.2.

Io emits twice as much heat as the Earth, due primarily to tidal forces from Jupiter and the other large satellites continuously stretching and compressing the interior of Io.

The satellite with the largest orbital period in the solar system is Sinope, at 758 days – more than 2 Earth years.

From Amalthea, Jupiter would cover 45° of the sky. This means that, when fully risen, Jupiter would stretch from the horizon to a point halfway up in Amalthea's sky.

As it sped past Jupiter, Pioneer 11 traveled faster than any other human-made object – over 172 200 km/hr.

Apart from Earth's Moon, the satellite with the greatest number of named features is Ganymede, with 165.

The smallest moon yet discovered in the solar system is Leda, with a diameter of only 8 km.

Early Jupiter photograph. Common, Andrew Ainslee, "Photographs of Jupiter" *Observatory*, **3** (no. 34): plate III, London, 1880. (Linda Hall Library)

The brightest natural satellite visible from Earth (not counting the Moon) is Ganymede, with an apparent visual magnitude of 4.4, at opposition. Io is next brightest at magnitude 4.7; then Europa at 5.1 and Callisto at 5.4.

Jupiter's average apparent motion (against the background of stars) is approximately 5 minutes of arc per day. Thus, in a little over six days Jupiter can move the width of the Full Moon.

"Mutual event seasons" of the Galilean satellites (the times when mutual phenomena can be observed) occur every 6 years, twice during each orbit of Jupiter around the Sun as the Earth crosses the orbital plane of the satellites.

Under favorable conditions, Jupiter can cast a visible shadow at night.

Daytime surface temperatures of the Galilean satellites are close to 100 K ($-173\,°C$).

Jupiter's gravitational effect on the Sun is such that the Sun's radial velocity changes by about 12.5 m/s with a period of about 12 years.

In 1675, the Dutch astronomer Ole Romer used eclipses of Jupiter's satellites to make the first determination of the speed of light. The value he arrived at was 225 300 km/s, a little slower than the modern value of 299 793 km/s.

The immense gravitational field of Jupiter causes gaps in the asteroid belt where an asteroid's orbital period forms a simple ratio with that of Jupiter. For example, gaps exist at the 1:2 and 1:3 resonance orbits.

Observing data

Future dates of conjunction

8 May 2000
14 Jun 2001
20 Jul 2002
22 Aug 2003
21 Sep 2004
22 Oct 2005
22 Nov 2006
23 Dec 2007
24 Jan 2009
28 Feb 2010

Future dates of opposition

23 Oct 1999
28 Nov 2000
1 Jan 2002
2 Feb 2003
4 Mar 2004
3 Apr 2005
4 May 2006
6 Jun 2007
9 Jul 2008
15 Aug 2009
21 Sep 2010

Future close conjunctions of Jupiter and the visible planets

	Close conjunctions	
	Date	Separation in declination
with Mercury		
	8 May 2000	52'
	20 Jul 2002	75'
	26 Jul 2003	23'
	28 Sep 2004	40'
	6 Oct 2005	88'
with Venus		
	17 May 2000	01'
	5 Aug 2001	72'
	3 Jun 2002	99'
	21 Aug 2003	34'
	4 Nov 2004	36'
	2 Sep 2005	82'
with Mars		
	6 Apr 2000	66'
	3 Jul 2002	49'
	27 Sep 2004	12'
with Saturn		
	31 May 2000	71'

Recent data   The most recent data comes from the Galileo spacecraft. Launched on 18 October 1989, Galileo was comprised of an orbiter and a probe which was released from the main spacecraft on 12 July 1995, and which entered the atmosphere of Jupiter on 7 December 1995. Surprising findings of the probe

Mars, Jupiter and Saturn. Hevelius, Johannes, *Selenographia: sive, Lunae descriptio*, Gdansk, 1647. (Linda Hall Library)

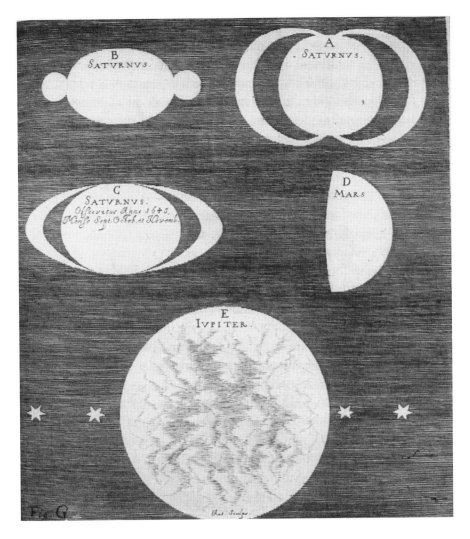

were the low amount of water within the atmosphere of Jupiter, much higher wind speeds than expected and low abundances of several elements.

The probe made the most difficult planetary atmospheric entry ever attempted. It survived entry speeds of over 170 590 km/hr, temperatures twice those on the surface of the Sun and deceleration forces up to 230 times the strength of gravity on Earth. It relayed data obtained during its 57-min descent mission back to the Galileo orbiter more than 209 200 km overhead for storage and transmission to Earth.

The probe detected extremely strong winds and very intense turbulence during its descent through Jupiter's thick atmosphere. This provides evidence that the energy source driving much of Jupiter's distinctive circulation phenomena is probably heat escaping from the deep interior of the planet. The probe also discovered an intense new radiation belt approximately 49 900 km above Jupiter's cloud tops.

225

Jupiter as seen by Voyager 2. (NASA)

Jupiter's atmosphere contains significantly lower than expected levels of helium, neon and certain heavy elements, such as carbon, oxygen and sulfur. It was widely held that Jupiter has a bulk composition similar to that of the gas and dust cloud of the primordial solar nebula from which the planets and our Sun were formed, with added heavy elements from comets and meteorites. The probe's measurements of helium and neon abundances cast some doubt on that theory. Some scientists are already re-evaluating their views of how Jupiter evolved from the solar nebula.

As the probe entered Jupiter's atmosphere, instruments showed the upper-level atmospheric density to be much greater than expected. Atmospheric temperatures were also much higher than predicted. The high temperatures cannot be explained by current theory. Apparently, there is something unexplained occurring in the upper Jovian atmosphere, causing higher-than-predicted temperatures to be found there.

Following probe parachute deployment, six science instruments on the probe collected data throughout 156 km of the descent. During that time, the probe endured severe winds, periods of intense cold and heat and strong turbulence. It eventually stopped transmitting when the temperature and pressure became too great for it to function as intended.

Earth-based telescopic observations suggest that the probe entry site may well have been one of the least cloudy areas on Jupiter. Researchers had hoped to study Jupiter's three distinct layers of clouds: a topmost layer of ammonia crystals, a middle layer of ammonium hydrosulfide and a final, thick layer of water and ice crystals. Unfortunately, the Galileo probe did not provide conclusive proof of such an atmospheric structure.

Some indication of a high-level ammonia ice cloud was detected by the net flux radiometer. Evidence for a thin cloud which might be the postulated ammonium hydrosulfide cloud was provided by another experiment. There was no data to suggest the presence of water clouds of any significance. The vertical temperature gradient obtained by the atmospheric structure instrument was characteristic of a dry atmosphere, free of condensation. Only the one, distinctive cloud structure was identified, and that was of modest proportion.

Due to Jupiter's vast size and rapid rotation rate, it was anticipated that the Galileo probe would find strong winds. The measured wind speeds surprised even veteran researchers. Instead of gusts of up to 322 km/hr, the probe recorded fairly constant winds at an amazing 531 km/hr. Wind speed did not vary significantly during the active transmission portion of the probe's trip through the Jovian atmosphere. This suggests that Jupiter's winds are not caused by differences in the intensity of sunlight between the equator and the poles. Nor are the winds products of heat released by water condensation, as on Earth. The origin of Jupiter's winds appears to be the internal heat source which radiates energy up into the atmosphere from the planet's deep interior. Because of this internal heat source, Jupiter's overall atmospheric movement is dominated by a jet-stream-like mechanism rather than swirling hurricane or tornado-like storms.

Since the Voyager spacecraft photographed lightning on the night side of Jupiter, astronomers were eager to measure the frequency with which such discharges occur. The Galileo probe found that lightning occurs on Jupiter only about one-tenth as often as on Earth. Such a discovery is consistent with the low percentage of water clouds. The small number of lightning discharges reduces the probability of finding complex organic molecules in Jupiter's atmosphere, particularly given its hostile, predominantly hydrogen composition.

Voyager 2 image of Io. (NASA)

Satellites

	Size (km)	Mass (kg)[a]	Density (g/cm^3)	Orbital period	Eccentricity	Inclination	Distance from planet (km)
Metis	50	—	—	07^{h}04.5^m	0.041	~0°	1.2796 × 10^5
Adrastea	30	—	—	07^{h}09.5^m	~0	~0°	1.2898 × 10^5
Amalthea	262(× 146 × 134)	—	—	11^{h}57.4^m	0.003	0.40°	1.813 × 10^5
Thebe	110	—	—	16^{h}11.3^m	0.0015	0.8°	2.219 × 10^5
Io	3630	8.94 × 10^{22}	3.57	01^{d}18^{h}27.6^m	0.041	0.040°	4.216 × 10^5
Europa	3120	4.799 × 10^{22}	3.018	03^{d}13^{h}14.6^m	0.0101	0.470°	6.709 × 10^5
Ganymede	5268	1.482 × 10^{23}	1.936	07^{d}03^{h}42.6^m	0.0015	0.195°	1.07 × 10^6
Callisto	4806	1.076 × 10^{23}	1.851	16^{d}16^{h}32.2^m	0.007	0.281°	1.883 × 10^6
Leda	8	—	—	238^{d}17^{h}16.8^m	0.163	27°	1.1094 × 10^7
Himalia	180	—	—	250^{d}13^{h}35.3^m	0.163	0.28°	1.148 × 10^7
Lysithea	24	—	—	259^{d}05^{h}16.8^m	0.107	29°	1.172 × 10^7
Elara	90	—	—	259^{d}15^{h}40^m	0.207	28°	1.1737 × 10^7
Ananke	20	—	—	631^d (r)[b]	0.169	147°	2.12 × 10^7
Carme	30	—	—	692^d (r)	0.207	163°	2.26 × 10^7
Pasiphae	36	—	—	735^d (r)	0.378	148°	2.35 × 10^7
Sinope	28	—	—	758^d (r)	0.275	153°	2.37 × 10^7

Notes:

[a] Satellites with no value for their mass have never had this quantity measured to any degree of accuracy.

[b] Retrograde.

Historical timeline	summer, 364 BC	The Chinese astronomer Gan De made a visual observation of what is almost certainly Ganymede, 1974 years before Galileo. At the time, Jupiter was within the boundaries of the constellation Aquarius.
	3 September 240 BC	δ Cancri was occulted by Jupiter. The event was recorded by Ptolemy.
	23 August 757	The Chinese observed and recorded an occultation of Jupiter by Venus.
	4 May 773	An observation of the occultation of β Scorpii by Jupiter was recorded in China; the Chinese also recorded occultations of this star by Jupiter on 15 May 1034, 17 April 1283 and 28 January 1366.
	7 January 1610	Italian astronomer Galileo Galilei first observed the four large moons of Jupiter.
	15 March 1611	Galileo reported Jupiter without a visible satellite for several hours. Calculations show that at least one satellite was visible at all times; his observations demonstrate the quality of his telescope.
	1626	Belgian astronomer Godfried Wendilin demonstrated that Kepler's laws are valid for the Galilean satellites of Jupiter.
	1664	British chemist and physicist Robert Hooke observed the rotation of Jupiter. He also discovered the Great Red Spot.
	1665	Italian-born, French astronomer Jean-Dominique Cassini measured the rotational rate of Jupiter.
	12 November 1681	William Molyneux became the first observer to see Jupiter without a visible satellite.
	1834	Marie-Charles-Théodore Damoiseau published *Ecliptical Tables of Jupiter's Satellites*.
	1866	US astronomer Daniel Kirkwood showed that the influence of Jupiter's gravity causes gaps in the distribution of the orbits of asteroids.
	23 June 1878	The first of 731 photometric observations of eclipses of Galilean satellites was made at the Harvard College Observatory.
	9 September 1892	US astronomer Edward Emerson Barnard discovered Amalthea.
	1904	US astronomer Charles Dillon Perrine discovered Himalia.
	1905	Charles Dillon Perrine discovered Elara.
	1908	P. Mellote discovered Pasiphae.
	1914	US astronomer Seth Barnes Nicholson discovered Sinope.
	1938	Seth Barnes Nicholson discovered Carme and Lysithea.
	1951	Seth Barnes Nicholson discovered Ananke.

Jupiter. *Annals of the Astronomical Observatory of Harvard College*, **8**, part 2, Cambridge, MA, 1876. (Michael E. Bakich collection)

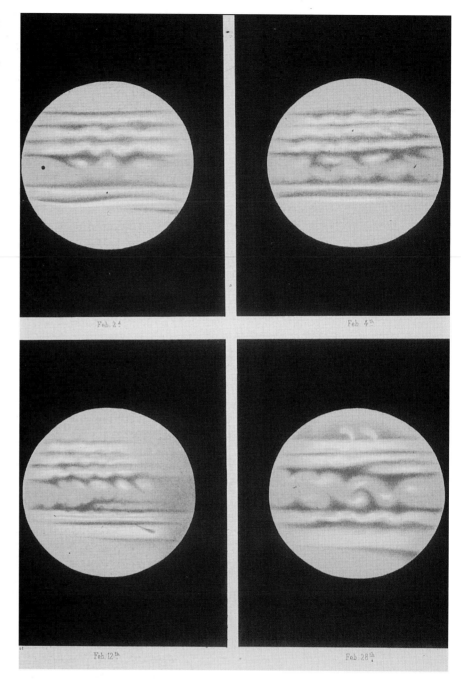

1955	US astronomer Kenneth Lynn Franklin detected radio emissions from Jupiter.
4 March 1969	A partial eclipse (passage through Jupiter's shadow) of Callisto was observed by German astronomer J. Classen at the Pulsnitz Observatory. The visual magnitude of Callisto dropped to 10.6, but the satellite never disappeared.
3 March 1972	US space probe Pioneer 10 launched toward Jupiter and Saturn; it passed within 132 000 km on 3 December 1973; returned 500 images of Jupiter and satellites; collected data on Jovian magnetic field, solar wind interactions and trapped charged particles.
7 June 1972	Ganymede occulted the 8th magnitude star SAO 186800.
1974	US astronomer Charles T. Kowal discovered Leda.
5 April 1974	US space probe Pioneer 11 launched toward Jupiter; passed within 42 800 km on 2 December 1974, on its way to Saturn; measured charged particles and magnetic field; returned photographs.
20 August 1977	US space probe Voyager 2 launched toward Jupiter (prior to the launch of Voyager 1); on 9 July 1979 it flew 650 000 km above the clouds of Jupiter, on its way toward the three other Jovian planets.
5 September 1977	US space probe Voyager 1 launched toward Jupiter; on 5 March 1979, it flew 349 010 km above the cloud tops of Jupiter, on its way to Saturn.
March 1979	Voyager 1 scientists discovered a thin ring around Jupiter.
8 March 1979	Voyager 1 returned the first photographs showing active volcanoes on a body other than Earth (Io).
18 October 1989	US–European space probe Galileo launched.
8 February 1992	US space probe Ulysses (launched 6 October 1990), on its way to study the Sun's poles, swung by Jupiter for a gravity boost.
9 April 1997	The spacecraft Galileo imaged fractures on Europa reminiscent of the disrupted pack ice often seen in Arctic seas on the Earth.
8 November 2001	From UT 16^h27^m to UT 16^h44^m, Earthbound observers will be able to see Jupiter without a visible satellite.

Saturn

Physical data

Size   120 536 km*

Mass 5.688×10^{26} kg

Escape velocity   35.49 km/s (127 764 km/hr)

Temperature range   Minimum   Average   Maximum
                              $-191\,°C$ $-130\,°C$ increases with depth

Oblateness   0.1076

Surface gravity   9.05 m/s² (equatorial)
                       10.14 m/s² (polar)[†]

Volume 8.183×10^{14} km³
              752 times that of Earth[‡]

Magnetic field strength and orientation   As is the case with Jupiter, the strong field is generated by Saturn's rotation – nearly as fast as Jupiter's – along with a similar interior of metallic liquid hydrogen.

Average surface field (tesla)	0.00002
Dipole moment (weber-meters)	3×10^{17}
Tilt from planetary axis	1°
Offset from planetary axis (planet radii)	0.0

Albedo (visual geometric albedo)   0.47

Density (water = 1)   0.69 g/cm³

Solar irradiance   15 watts/m²

Atmospheric pressure   varies with depth, >10 132 500 pascals

Composition of atmosphere		
	Molecular hydrogen (H_2)	>93%
	Helium (He)	>5%
	Methane (CH_4)	0.2%
	Water vapor (H_2O)	0.1%
	Ammonia (NH_3)	0.01%
	Ethane (C_2H_6)	0.0005%
	Phosphine (PH_3)	0.0001
	Hydrogen sulfide (H_2S)	<0.0001
	Methylamine (CH_3NH_2)	<0.0001%
	Acetylene (C_2H_2)	0.00001%
	Monodeuteromethane (CH_3D)	0.000002%
	(*continued opposite*)	

* Polar diameter 107 566 km.
[†] The rapid rotation of Saturn causes the planet to be flattened.
[‡] If Saturn were spherical, its volume would be 9.171×10^{14} km³, which would be 844 times that of Earth (using the equatorial diameter).

Saturn. Herschel, William, "Observations of a quintuple belt on the planet Saturn" *Philosophical Transactions*, **84**:28–32, London, 1794. (Linda Hall Library)

Compositions of atmosphere (cont.)	Hydrogen cyanide (HCN)	0.0000001%
	Germane (GeH$_4$)	<0.0000001%
	Hydrazine (N$_2$H$_4$)	<0.0000001%
	Ethylene (C$_2$H$_4$)	<0.0000001%
	Carbon monoxide (CO)	<0.00000001%

Maximum wind speeds   1440 km/hr

Outstanding cloud features

The atmosphere of Saturn is very similar to that of Jupiter. Although the cloud bands are not as pronounced, the nomenclature used to differentiate them is the same. Light colored cloud bands are called "zones" and darker bands are known as "belts." Colors are subtle, ranging from yellow and tan to light

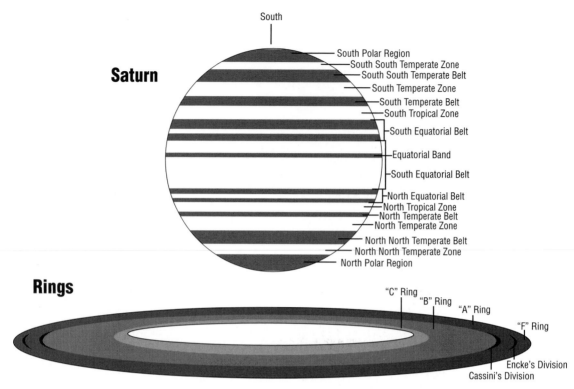

Typical observing nomenclature for Saturn. (Holley Bakich)

brown, depending on the observer. Temporary whitish spots are sometimes seen. These tend to last several months, on average.

The diagram shows the usual labeling of Saturn's cloud bands.

Orbital data

Period of rotation
0.42637 days
10.233 hours
$0^d10^h14^m$

Period of revolution (sidereal orbital period)
29.458 years
10759.3 days
$29^y167^d6.7^h$

Synodic period $378^d02^h09.6^m$

Equatorial velocity of rotation 37004.9 km/hr

Velocity of revolution 9.67 km/s (34812 km/hr)

Distance from Sun

Average	9.5388 AU	
	1429400000 km	
Maximum	10.072 AU	
	1506750000 km	

	Minimum	9.010 AU
		1 347 877 000 km
Distance from Earth	Maximum	11.09 AU
		1 658 854 980 km
	Minimum	7.99 AU
		1 195 772 020 km

Apparent size of Sun (average)   0.056°

Apparent brightness of Sun $m_{vis} = -21.8$

Orbital eccentricity   0.0560

Orbital inclination   2.488°

Inclination of equator to orbit   25.33°

Observational data

Maximum angular distance from Sun   180°

Brilliancy at opposition	Maximum	−0.3
	Minimum	+0.9

Angular size*	Maximum	20.75″
	Minimum	18.44″

Early ideas

To the Sumerians, the planet Saturn was known as Ubu-idim or Sag-us. For the early Chinese, it was Tu xing. In Babylon, it was the star of the god Genna, Ninib or Nin-urta. Ninib was primarily an agricultural deity, regulating the changes of seasons. The Babylonians did not set the planets up as equivalent to the gods. Instead, they saw them as celestial talismans of the gods' power and intentions. In essence, the planet became the property of the individual deity to which it was linked.

The Babylonians watched Saturn closely, and recorded its movements. They were particularly interested in the dates of five Saturnian events: the instants when the planet starts and ends its retrogression; the first visible heliacal rising; the last visible heliacal setting; and opposition. They also computed the longitudinal position of the planet on the ecliptic for these dates.

Through careful observation, the Mesopotamians saw that Saturn takes almost 30 years to pass completely through the band of the zodiac, making it the slowest-moving of the planets. Because of this fact, they correctly surmised that it was the most distant of the planets.

The Greeks first called Saturn "the star of Kronos." As the planet Saturn, Kronos was recognized to be a deposed king. He had once ruled the world as the Sun rules the day. For this reason they called it the "sun of the night" and

* This measurement is the apparent angular equatorial diameter of Saturn, measured in seconds of arc, as seen from Earth. Note: Saturn's polar diameter is 89% of its equatorial diameter.

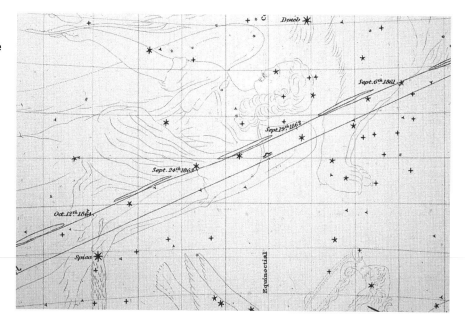

Saturn's path for several years, showing retrograde loops. Proctor, Richard Anthony, *Saturn and its System*, London, 1865. (Linda Hall Library)

in some ways made it the Sun's antithesis. And, although it may seem strange to us, the Greeks occasionally called Saturn Phainon, or "brilliant star." Although not nearly the brightest planet, if found in an area where there are no bright stars, Saturn can stand out quite nicely in a dark night sky.

The ancient Greeks (and later the Romans) closely associated the planets with their gods. Saturn was the slowest-moving, most distant of the planets known at that time. Therefore the Greek god Kronos (and his Roman counterpart) was believed to be lethargic and have a dark temperament. His temperament and his Roman name gave us the word *saturnine*, meaning uncommunicative, melancholy and slow.

The planet also gave its name to the last day of the week. The Romans called it "Saturnus dies," which translates as "Saturn's day." In Rome Saturn's seven-day feast was the Saturnalia, a holiday that began on December 17th and was timed in part by the winter solstice – the day of shortest daylight in the Northern Hemisphere.

Early astrology associated Saturn with lead, a ponderous metal that wedded naturally with the way the planet inched along the sky. The planet signaled trouble, trial, failure and doom. According to Ptolemy each planet rules specific signs of the zodiac, and Saturn governs Aquarius and Capricorn. This connection between Saturn and one of the signs is clear. During the time of Ptolemy, the winter solstice occurred when the Sun was in the constellation Capricornus, which meant that the signs and stars ruled by Saturn belonged to the darkest, coldest season.

Beginning with Galileo's telescopic observations of Saturn, the story of Saturn is inextricably bound to its magnificent system of rings. The initial question, "What were they?" was answered by Huygens. Physical attributes were supplied by Cassini, Herschel, Encke and many others. For the nature

Saturn. Smyth, William Henry, *The Cycle of Celestial Objects*, London, 1860. (Michael E. Bakich collection)

of the rings we look to Maxwell, Roche, Barnard, Keeler, Campbell, etc. For a more detailed discussion of the rings, see the following section.

Permit me to indulge in a brief aside. In my search for illustrations to be included in this book, I had access to nearly 400 years of historical images, thanks to the Linda Hall Library. Saturn dominates these.* It seemed that every early observer who ever used a telescope drew Saturn, and drew it a lot. I suppose I should not have been surprised. Having spent thousands of hours at the eyepieces of telescopes of all sizes, I am well acquainted with the splendor of Saturn. Also, anyone who has ever conducted a public observing session cannot help but be pleased by the reaction of viewers of Saturn for the first time. It seems that the old adage is true – whatever their scientific prowess, humans recognize three astronomical objects: the Moon, Halley's Comet and the rings of Saturn.

Important concepts

The rings

Before we launch into a detailed discussion of the rings, a short chronology is in order. The first to observe Saturn's rings was the Italian astronomer Galileo Galilei, in 1610, but it was not until 1655 that the true nature of the rings was discerned. In that year, the Dutch astronomer Christiaan Huygens stated that Saturn was surrounded by a solid ring. Four years later, Huygens detailed the geometry of the Earth passing through the plane of Saturn's ring. In the same year, British physicist James Clerk Maxwell stated that Saturn's ring would be broken by the tensions of attraction and centrifugal force; he suggested that Saturn's ring consisted of numerous small bodies which, like a ring of meteorites, freely circle the planet.

In 1676, the French astronomer Jean-Dominique Cassini discovered a gap between the outer (A-) and the inner (B-) rings. This soon became known as

---

* Except for a brief time period at the end of the nineteenth and beginning of the twentieth century when "Mars mania" was rampant.

An image of Saturn from the man who correctly identified the nature of the rings. Huygens, Christiaan, *Systema Saturnium*, The Hague, 1659. (Linda Hall Library)

Saturn, showing the Cassini Division as a dark area. Cassini, Jean-Dominique, "An extract of Signor Cassini's letter ..." *Philosophical Transactions*, **11**:689–90, London, 1676. (Linda Hall Library)

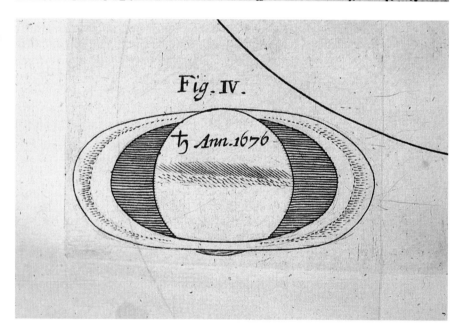

the Cassini Division. In 1837, the German astronomer Johann Encke noticed a dark band – later called Encke's Division* – in the middle of the A-ring. In 1849, Edouard Roche proposed a ring formation theory. He stated that a fluid satellite had approached Saturn so closely that tidal forces had broken it apart.

* Also known as Encke's Gap.

Jean-Dominique
Cassini. (Michael E.
Bakich collection)

During the mid-nineteenth century, more and more evidence was being gathered that the rings of Saturn were not solid, but made of numerous particles. In 1883, A. A. Common took the first photograph of the rings. Late in the nineteenth century, measurements showed that the inner part of the ring system rotated more rapidly than the outer part – observational evidence of non-solid rings.

Some of the
hundreds of rings
of Saturn. (NASA)

With more detailed observations of Saturn's rings being made, theories of their formation seemed to know no end. Today, three prevailing theories are under consideration. Before we discuss the theories, however, a short review of Roche's limit may prove helpful.

The Roche limit is the minimum distance from the center of a planet that a satellite can maintain equilibrium, that is, without being pulled apart by tidal forces. If a planet and a moon have the same density, the Roche limit is 2.446 times the radius of the planet. Any sizable satellite orbiting inside the Roche limit will be destroyed. The Roche limit for the Earth is 18470 km. If our Moon somehow ventured within this distance, it would be pulled apart by tidal forces and a ring system might form around the Earth. The ring systems of the four Jovian planets are within their respective Roche limit. Listed are the approximate values:

Jupiter	173800 km
Saturn	148000 km
Uranus	62800 km
Neptune	59500 km

One theory of the formation of Saturn's rings assumes that the particles are material left over from the formation of the solar system. This matter could not form into a satellite because it was located within the Roche limit for Saturn. Eventually, collisions and interactions formed the rings as we see them today.

Another theory states that at some time early in its history, one of Saturn's moons ventured too close to the planet. Because this satellite was inside the

Roche limit, it was pulled apart by tidal forces. As above, after a sufficient length of time the ring system formed.

The third theory is that of violent encounter. At an early time in the history of the planet and its moons, one of Saturn's satellites suffered a devastating impact, either from bombardment by meteors or through a collision with another moon. Astronomers view these three theories as equally probable.

When astronomers gained some grasp of the origin of Saturn's rings, their next question was, "What keeps them in place?" Most theoretical calculations show the rings either being pulled into Saturn, due to the planet's gravity, or dissipating in space, due to collisions among ring particles.

The leading theory which explains how the ring particles are kept in place is the theory of shepherding moons. The gravitational attraction of these moons forces the particles into rings that rotate around Saturn.

Shepherding moons are satellites that orbit near a ring. Due to gravitational effects from the shepherding moon, the edges of the rings are kept sharp and distinct. If the shepherding moon was not present, then the ring material would have a tendency to spread out. In the case of two satellites orbiting on each side of a ring, the ring will be constrained on both sides into a narrow band.

The concept of shepherding moons was first proposed by Peter Goldreich and Scott Tremaine in 1979 to explain why the Uranian rings were so narrow. Voyager 1 scientists discovered the first Saturnian pair of shepherding moons, Prometheus and Pandora, in 1980, near the narrow F-ring. (Voyager 2 later found shepherding moons at Uranus in 1986.) During the Saturn ring-plane crossing of 22 May 1995, the Hubble Space Telescope discovered a third shepherding moon orbiting near the F-ring.

The rings have a maximum thickness of 0.8 km, but are probably much thinner. Results from Voyager 2 indicate an average thickness of no more than about 200 m. Space probes have shown that the large rings are actually made up of a series of smaller ringlets, giving the rings a highly structured appearance. Voyager scientists also discovered the presence of wave patterns in the rings.

Near-infrared observations from Earth have shown that the surface of the ring particles is predominantly water ice. Impurities have been detected, suggesting some small amount of silicate material may be mixed in with the ring particles.

The size of most of the ring particles ranges from 1 cm to 5 m. It is likely a few kilometer-sized objects exist as well. One of Saturn's moons, Pan, is inside the Encke Division and is 20 km in diameter. Additional moonlets may yet be discovered in some of the other gaps in the rings.

Voyager also found strange dark radial features up to 20 000 km in length that move about in curious patterns within the B-ring. Since the spokes have been observed on both sides of the ring plane, they are thought to be microscopic grains that have become charged and are levitating away from the ring plane. Since the spokes are seen to rotate at the same rate as Saturn's magnetic field, it is thought that the electromagnetic forces are also at work.

Voyager 2 image of
Saturn's rings
showing the spoke-
like structure.
(NASA)

Name	Distance from center of ring to Saturn (km)	Width (km)
D-ring	71 000	7500
C-ring	83 500	17 500
Maxwell Division	88 000	270
B-ring	105 200	25 500
Cassini Division	120 000	4200
A-ring	129 500	14 600
Encke Division	133 500	325
Keeler Division	137 000	35
F-ring	140 600	30–500
G-ring	170 500	8000
E-ring	250 000	300 000

The Cassini Division is the largest clear space in the ring system. Voyager 2 revealed that this space is not empty, but the majority of the ring debris in that area has been removed. The cause is believed to be the moon Mimas. The Cassini Division is maintained by a 2:1 resonance with Mimas. According to Kepler's third law, any particle that orbits at the distance of Cassini's Division would orbit exactly two times every time Mimas completes one orbit. Thus a particle in the gap would feel a gravitational attraction from Mimas every other time it completed an orbit. Eventually, Mimas would pull the particle out of its orbit, leaving a gap behind.

Saturn's orbit, showing ring orientations. Huygens, Christiaan, *Systema Saturnium*, The Hague, 1659. (Linda Hall Library)

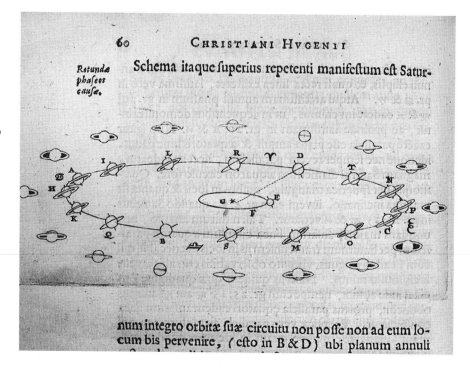

Ring-plane crossings

It takes Saturn approximately 29½ years to complete one revolution around the Sun. During this time, the angle of Saturn's rings relative to the Sun varies by 27.3°, which is the tilt of the rings to Saturn's orbital plane. Twice during Saturn's orbit, the rings are edge-on to the Sun. Since, as seen from Saturn, the Earth appears not more then 6° from the Sun, it too crosses the ring plane at roughly the same time. Due to the relative thinness of the rings compared to their distance from Earth, they seem to disappear in all but the largest telescopes.

The Earth may experience either one or three ring-plane crossings (always an odd number) during any half-orbit of Saturn. If there is only one ring-plane crossing then Saturn and the Earth will be on almost opposite sides of the Sun, making observations of Saturn difficult. If there are three crossings, the middle one is near opposition and the other two are near quadratures. The chance of three intersections is about 53% and the chance of one intersection is about 47%. There is an occasional case where the Earth hangs in the plane without passing through it.

When the rings of Saturn are nearly edge-on to Earth, the glare from the rings is reduced considerably, and faint objects near Saturn are easier to see. Months before and after the ring-plane crossings, unique observations of Saturn, its rings and moons can be made from Earth which are available at no other time. What can be learned during a ring-plane crossing? Here is a partial list:

(1) Thirteen of Saturn's moons have been discovered around the time of a ring-plane crossing.
(2) A large number of eclipses and occultations of the satellites by Saturn will occur, and accurate timings of these events can be obtained.
(3) The cloud tops of Saturn are more easily observed without the glare from the rings.
(4) Direct observations of the thickness of the main rings can be performed.
(5) Observations of the dark side of the rings and of the C-ring are possible during a ring-plane crossing.
(6) The small inner moons of Saturn can be observed with better precision.
(7) The tenuous E- and G-rings may be observed from Earth, which are normally obscured by the glare from Saturn's rings. The faint, outermost E-ring is easier to detect when viewed edge-on due to the greater amount of material in the line-of-sight.
(8) Information obtained during the 1995–1996 ring-plane crossing was invaluable for the Cassini mission to Saturn. Cassini will actually pass through the outer rings of Saturn during its orbit insertion – passing between the F- and G-rings at a distance of 2.67 Saturn radii from the planet. Observations of the G-ring can help assess the relative risk of ring particle impact to Cassini before it arrives at Saturn.

The next ring-plane crossing will be the hard-to-observe single crossing which will occur on 4 September 2009. Another single crossing will take place on 23 March 2025. The next triple ring-plane crossing does not occur until 2038.

Interesting facts

From Saturn, the Sun appears 1% as bright as from Earth.

Of all solar system objects whose densities have been measured, the least dense is Prometheus. The density of this satellite is only slightly more than a quarter the density of water.

Saturn is 95 times as massive as the Earth.

The most reflective object in the solar system is Enceladus, with a geometric albedo of 1.0. This means that Enceladus reflects essentially 100% of the light striking it back into space.

The diameter of the rings of Saturn is 225% that of the equatorial diameter of the planet.

Titan is the only body in the solar system, apart from Earth, with a primarily nitrogen atmosphere.

Saturn is the only major solar system body whose density is less than water. If you could find an ocean big enough, Saturn would float.

Daytime temperatures on the surfaces of Saturn's satellites average about 70 K ($-203\,°$C).

Saturn's north pole is pointed toward right ascension 02^{h}34^m and declination $+83.3°$ (1950.0 coordinates). This is a point within the boundaries of the constellation Cepheus. It is interesting that Saturn's north celestial pole is only 6° from that of Earth.

Thirteen of Saturn's moons have been discovered around the time of a ring-plane crossing.

Saturn's polar diameter is only 89% of its equatorial diameter.

Saturn's average apparent motion (against the background of stars) is approximately 2 minutes of arc per day. Thus, in about 15 days, Saturn can move the width of the Full Moon.

The nature of Saturn's rings was first identified in 1655, by the Dutch astronomer Christiaan Huygens.

With respect to the orbit of Saturn, the rings are tilted 27.3°.

844 Earths could fit inside Saturn, if Saturn were spherical.

For about a week before and after each Voyager encounter with Saturn, lightning was detected on the planet. The discharges were associated with a long-lived atmospheric lightning storm or system of storms just north of Saturn's equator and extending 60° (64 400 km) in longitude.

The orientation of Saturn's rings at opposition can dramatically increase its brightness. The difference between the rings being totally open and Earth passing through the plane of the rings can be as much as 0.9 magnitude.

Observing data   **Future dates of conjunction**

10 May 2000
25 May 2001
9 Jun 2002
24 Jun 2003
8 Jul 2004
23 Jul 2005
7 Aug 2006
22 Aug 2007
4 Sep 2008
17 Sep 2009
1 Oct 2010

Future dates of opposition

6 Nov 1999
19 Nov 2000
3 Dec 2001
17 Dec 2002
31 Dec 2003
13 Jan 2005
28 Jan 2006
10 Feb 2007
24 Feb 2008
8 Mar 2009
22 Mar 2010

Future close conjunctions of Saturn and the visible planets

	Close conjunctions	
	Date	Separation in declination
with Mercury		
	2 Jul 2002	14'
	1 Jul 2003	93'
	26 Jun 2006	85'
with Venus		
	18 May 2000	74'
	15 Jul 2001	44'
	8 Jul 2003	49'
	25 Jun 2005	78'
with Mars		
	24 May 2004	95'
with Jupiter		
	31 May 2000	71'

Recent data

At the time of writing, the bulk of the recent data on Saturn has come from Voyagers 1 and 2. These spacecraft encountered Saturn nine months apart, in November 1980 and August 1981.

Voyager 1 found that about 7 percent of the volume of Saturn's upper atmosphere is helium. The same measurement at Jupiter found 11 percent. For both of these planets, the remaining gas is almost all hydrogen. Since Saturn's internal helium abundance was expected to be the same as Jupiter's, the lower abundance of helium in the upper atmosphere may imply that the heavier helium may be slowly sinking through Saturn's hydrogen. Such a process might explain the excess heat that Saturn radiates over energy it receives from the Sun.

Voyager 2 image of
Saturn. (NASA)

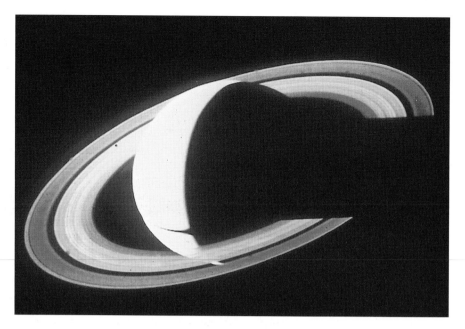

Near the equator, the Voyagers measured wind speeds of 1440 km/hr (400 m/s). The wind blows mostly in an easterly direction. Strongest winds are found near the equator, and velocity falls off uniformly at higher latitudes. The discovery of dominant eastward jet streams indicates that winds are not confined to the cloud layer, but must extend inward at least 1950 km.

Voyager 2 measured Saturn's minimum temperature of 82 K. This reading was found where the atmospheric pressure equals 7093 pascals. The temperature increased to 143 K where the atmospheric pressure was approximately 121 590 pascals. This was the deepest level probed by either spacecraft.

The Voyagers found aurora-like ultraviolet emissions of hydrogen at mid-latitudes in the atmosphere, and auroras at polar latitudes (above 65°). The high-level auroral activity may lead to formation of complex hydrocarbon molecules that are carried toward the equator.

As might be imagined, Saturn's rings were studied in great detail. Voyager 2's photopolarimeter measured changes in starlight from the star δ Scorpii as Voyager 2 flew above the rings and the light passed through them.

The star-occultation experiment showed that few clear gaps exist in the rings. The structure in the B-ring appears to be variations in density waves or other, stationary, forms of waves. Density waves are formed by the gravitational effects of Saturn's satellites. The small-scale structure of the rings may therefore be transitory, although larger-scale features, such as the Cassini and Encke Divisions, appear more permanent.

The edges of the rings where the few gaps exist are so sharp that the ring must be less than about 200 m thick there, and may be only 10 m thick.

In almost every case where clear gaps do appear in the rings, eccentric ringlets are found. All show variations in brightness. Some differences are due to the existence of clumps or kinks, and others to nearly complete absence

Saturn, three views. *Annals of the Astronomical Observatory of Harvard College*, **2**, part 2, Cambridge, MA, 1867. (Michael E. Bakich collection)

Fig. 61.   November 15, 1850.

Fig. 62.   January 7, 1851.

Fig. 65.   October 20, 1851.

Saturn's F-ring. Note the presence of two shepherding satellites. (NASA)

of material. Some scientists believe the only plausible explanation for the clear regions and kinky ringlets is the presence of nearby undetected satellites.

Saturn's F-ring was discovered by Pioneer 11 in 1979. Photographs of the F-ring taken by Voyager 1 showed three separate strands that appear twisted or braided. At higher resolution, Voyager 2 found five separate strands in a region that had no apparent braiding and, surprisingly, revealed only one small region where the F-ring appeared twisted. The twists are believed to originate in gravitational perturbations caused by one of the two shepherding satellites, Prometheus.

Both Voyagers recorded the appearance of a spoke-like structure in the rings. As the spacecraft approached Saturn, the spokes appeared dark against a bright ring background. As the Voyagers departed, the spokes appeared brighter than the surrounding ring areas, indicating that the material scatters reflected sunlight more efficiently in a forward direction, a quality that is characteristic of fine, dust-sized particles.

After the fabulous discoveries related to the satellites of Jupiter, Voyager project scientists eagerly anticipated the data that would be returned from the moons of Saturn. This was especially true of Titan. Titan is the second largest satellite in the solar system, and the only one known to have a dense atmosphere. It has long been held that the chemistry in Titan's atmosphere may be similar to that of the Earth several billion years ago.

Titan's surface cannot be seen in any Voyager photographs; it is hidden by a dense, photochemical haze whose main layer is about 320 km above Titan's

Saturn, with eight satellites. *Annals of the Astronomical Observatory of Harvard College*, **2**, part 2, Cambridge, MA, 1867. (Michael E. Bakich collection)

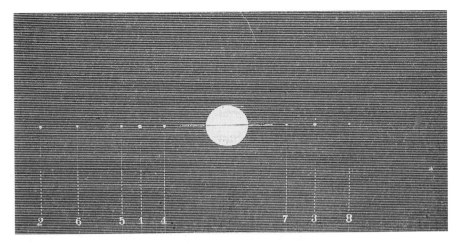

surface. Several distinct, detached haze layers can be seen above the opaque haze layer. The haze layers merge with the main layer over the north pole of Titan, forming a dark ring around the pole.

Atmospheric pressure near Titan's surface is about 162 120 pascals, 60 percent greater than that of Earth. The atmosphere is made up almost entirely of nitrogen (N_2), with traces of methane, ethane, acetylene, propane, diacetylene, methylacetylene, hydrogen cyanide, cyanoacetylene, cyanogen, carbon dioxide and carbon monoxide.

The surface temperature appears to be about 95 K, only 4° above the triple-point temperature of methane. Methane, however, appears to be below its saturation pressure near Titan's surface. This means that rivers and lakes of methane probably could not exist, in spite of the tantalizing analogy to water on Earth. On the other hand, scientists believe lakes of ethane exist, and some methane is probably dissolved in the ethane.

Titan has no intrinsic magnetic field, however interaction with Saturn's magnetosphere creates a magnetic wake behind Titan. The big satellite also serves as a source for both neutral and charged hydrogen atoms in Saturn's magnetosphere.

Mimas, Enceladus, Tethys, Dione and Rhea are approximately spherical in shape and appear to be composed mostly of water ice. Enceladus reflects almost 100 percent of the sunlight that strikes it. All five satellites represent a size range that had not been explored before.

Enceladus appears to have by far the most active surface of any satellite in the system. At least five types of terrain have been identified on Enceladus. Although craters can be seen across portions of its surface, the lack of craters in other areas implies an age less than a few hundred million years for the youngest regions. It seems likely that parts of the surface are still undergoing change, since some areas are covered by ridged plains with no evidence of cratering down to the limit of resolution of Voyager 2's cameras, which was 2 km. Because it reflects so much sunlight, Enceladus's current surface temperature is only 72 K.

In appearance, Mimas is one of the most impressive objects seen by the Voyager cameras. Photographs show an immense impact crater one-third the diameter of Mimas itself. The crater, named Herschel, is 130 km wide. Herschel is 10 km deep, with a central mountain 8 km high.

Photographs of Tethys taken by Voyager 2 show an even larger impact crater, named Odysseus, nearly one-third the diameter of Tethys and larger than Mimas. In contrast to Mimas's Herschel, the floor of Odysseus returned to about the original shape of the surface, probably as a result of Tethys's larger gravity and the relative fluidity of water ice. A gigantic fracture covers three-quarters of Tethys's circumference. The fissure is about the size scientists would predict if Tethys were once fluid and its crust hardened before the interior, although the expansion of the interior due to freezing would not be expected to cause only one large crack. The canyon has been named Ithaca Chasma.

The Voyager spacecraft found an interesting relationship between Hyperion and Titan, caused by the irregular shape of Hyperion. Each time Hyperion passes Titan, the gravity of Titan gives Hyperion a tug and it tumbles erratically, changing orientation. In addition, the surface of Hyperion appears to be the oldest in the Saturnian system.

Voyager 2 photographed Phoebe after passing Saturn. Interestingly, Phoebe rotates on its axis about once in nine hours. Thus, unlike the other Saturnian satellites (except Hyperion), it does not always show the same face to the planet. If, as scientists believe, Phoebe is a captured asteroid with its composition unmodified since its formation in the outer solar system, it is the first such object that has been photographed at close enough range to show shape and surface brightness.

Like the rest of the planets, the size of Saturn's magnetosphere is determined by external pressure of the solar wind, which can be quite variable. When Voyager 2 entered the magnetosphere, the pressure from the solar wind was high and the magnetosphere extended only 1 100 000 km in the Sun's direction. Three days later, as Voyager 2 was outbound from Saturn, the magnetosphere was 70% larger. Unlike the other planets whose magnetic fields have been measured, Saturn's field is tipped less than 1° relative to the rotation poles.

In December 1996, Claude and François Roddier of the University of Hawaii, using the 3.6-m Canada–France–Hawaii telescope with adaptive optics, reported observations of 12 objects, potentially moonlets, orbiting within Saturn's F-ring. Each was approximately 20–40 km in diameter. At the time of writing they are awaiting confirmation.

Much new information will be forthcoming – if all goes according to plan – in 2004, when the Cassini spacecraft encounters Saturn.

Saturn and its
satellites.
Doppelmayr,
Johann Gabriel,
Atlas Coelestis,
Nuremberg, 1742.
(Linda Hall Library)

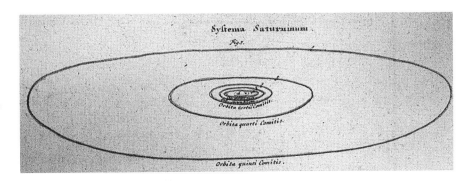

Satellites

	Size (km)	Mass (kg)[a]	Density (g/cm³)	Orbital period	Eccentricity	Inclination	Distance from planet (km)
Pan	20	—	—	13^h48^m	~0	~0°	1.33583×10^5
Atlas	37($\times 34.4 \times 27$)	—	—	$14^h26.7^m$	~0	~0°	1.3764×10^5
Prometheus	1528	1.4×10^{17}	0.270	$14^h42.7^m$	0.0042	0.0°	1.3935×10^5
Pandora	110($\times 88 \times 62$)	1.3×10^{17}	0.420	$15^h05.5^m$	0.0042	0.0°	1.417×10^5
Hyperion	370($\times 280 \times 226$)	—	—	$21^d06^h38.3^m$	0.1042	0.43°	1.4811×10^5
Epimetheus	138($\times 110 \times 110$)	5.05×10^{17}	0.630	$16^h40.2^m$	0.009	0.34°	1.51422×10^5
Mimas	398	3.75×10^{19}	1.140	$22^h37.1^m$	0.0202	1.53°	1.8552×10^5
Enceladus	498	7.3×10^{19}	1.120	$01^d08^h53.1^m$	0.0045	0.02°	2.3802×10^5
Calypso	30($\times 16 \times 16$)	—	—	$01^d21^h18.4^m$	~0	~0°	2.9466×10^5
Telesto	30($\times 25 \times 15$)	—	—	$01^d21^h18.4^m$	~0	~0°	2.9466×10^5
Tethys	1060	6.22×10^{20}	1.000	$01^d21^h18.4^m$	0	1.09°	2.9466×10^5
Dione	1120	1.052×10^{21}	1.440	$02^d17^h41.2^m$	0.0022	0.02°	3.774×10^5
Helene	32	—	—	$02^d17^h41.2^m$	0.005	0.2°	3.774×10^5
Rhea	1530	2.31×10^{21}	1.240	$04^d12^h25.2^m$	0.001	0.35°	5.2704×10^5
Titan	5150	1.3455×10^{23}	1.881	$21^d06^h38.3^m$	0.0292	0.33°	1.22185×10^6
Janus	199($\times 191 \times 151$)	4.98×10^{18}	0.650	$16^h40.2^m$	0.007	0.14°	1.51472×10^6
Iapetus	1436	1.59×10^{21}	1.020	$79^d07^h55.5^m$	0.0283	7.52°	3.5613×10^6
Phoebe	230($\times 220 \times 210$)	—	—	$550^d11^h31.2^m(r)$[b]	0.163	175.3°	1.2952×10^7

Notes:

[a] Satellites with no value for their mass have never had this quantity measured to any degree of accuracy.

[b] Retrograde.

Christiaan
Huygens. (Michael
E. Bakich
collection)

Historical
timeline

29 July 1027	The Chinese observed and recorded an occultation of Saturn by Mars.
1610	Galileo Galilei became the first to observe Saturn's rings; he thought the rings were "handles" or moons on either side of the planet. He said "I have observed the highest planet [Saturn] to be triple-bodied. This is to say that to my very great amazement Saturn was seen to me to be not a single star, but three together, which almost touch each other."
1612	Galileo was astounded when he found that the rings he first observed a couple of years earlier had disappeared; he was the first person to observe a Saturn ring-plane crossing.
1655	Dutch astronomer Christiaan Huygens proposed that Saturn was surrounded by a solid ring, "a thin, flat ring, nowhere touching, and inclined to the ecliptic."

25 March 1655	Using a refractor which provided a magnification of 50 (that he designed himself), Huygens discovered the first Saturn moon, Titan.
1656	Polish astronomer Johannes Hevelius postulated that Saturn's rings were two crescents attached to an ellipsoidal central body.
1658	British architect and astronomer Christopher Wren argued that an elliptical corona was attached to Saturn, and the planet and corona rotated about the major axis of the corona. He speculated that this corona was so thin that it was invisible when it was edge-on from Earth's perspective.
1659	Christiaan Huygens published his book, *Systema Saturnium*, in which he explains that every 14 to 15 years the Earth passes through the plane of Saturn's rings.
	British physicist James Clerk Maxwell stated that Saturn's ring would be broken by the tensions of attraction and centrifugal force; he suggested that Saturn's rings consist of numerous small bodies which, like a ring of meteorites, freely circle the planet.
1660	Jean Chapelain suggested that Saturn's rings are made up of a large number of very small satellites.
1664	Giuseppe Campani observed that the outer half of Saturn's ring is less bright than the inner half, but failed to recognize this as being two separate rings.
1671	Italian-born, French astronomer Jean-Dominique Cassini, during a ring-plane crossing by Earth, discovered Iapetus. He correctly postulated that Iapetus had light and dark sides, and that it always kept the same face turned to Saturn.
1672	Cassini discovered Rhea.
1676	Cassini discovered a gap in the rings which would later be named the Cassini Division. The outer ring would be called the A-ring and the brighter inner ring would be called the B-ring.
1684	Cassini discovered two more Saturn moons, Tethys and Dione, just prior to the ring-plane crossing of 1685. It was over 100 years before the next Saturn moon was discovered.
1780	German-born British astronomer Sir William Herschel reported seeing a "black list," or linear markings on one side of the A-ring near its inner edge. He had probably observed what would later be called the Encke Division.

Jupiter and Saturn
with the Cassini
Division. One of the
first images to
show this feature
as a true division.
Wright, Thomas,
*An Original Theory
or New Hypothesis
of the Universe*,
London, 1750.
(Linda Hall Library)

1787	Pierre Simon, marquis de Laplace suggested that Saturn had a large number of solid rings.
	Herschel suspected he had observed a new moon of Saturn (which would later turn out to be Enceladus), but waited until the next ring-plane crossing in 1789 to pursue it further.
1789	Sir William Herschel suggested that Saturn is surrounded by two solid rings. Herschel also discovered two new Saturn moons, Enceladus and Mimas, during the ring-plane crossing of 1789–1790. Herschel also found that Saturn is flattened at its poles, something he had suspected since 1776. He also made one of the earliest estimates of the thickness of the rings at 483 km, and reported observations of eclipses of Saturn's moons by the planet's shadow.
1790	Herschel determined the rotation period of Saturn's ring to be 10 hours 32 minutes.
1835	Friedrich Wilhelm Bessel was able to determine the orientation of Saturn's pole with unprecedented accuracy and found the precession period of the pole to be 340 000 years. (The modern estimate is 1 700 000 years.)
1837	The German astronomer Johann Encke observed a dark band in the middle of the A-ring. This band would be later known as Encke's Division, even though Encke never really saw it as a gap in the rings.
1848	William Lassell discovered Hyperion, during the ring-plane crossing of 1848–1849.
1849	Edouard Roche suggested that Saturn's ring system was formed when a fluid satellite had approached Saturn so closely that it had been torn apart by tidal forces.
1850	William Bond and George Bond observed a dark band across Saturn immediately adjacent to the interior edge of the B-ring. This ring was originally known as the crepe ring, and later officially became the C-ring. George Bond concluded that a system of narrow solid rings could not be stable and that Saturn's rings had to be fluid.
1852	Several observers noticed that the limbs of Saturn are visible through the C-ring. This observation made it difficult to defend the theory that the rings were solid.
1856	James Clerk Maxwell deduced that the Saturn rings cannot be solid and must be made of "an indefinite number of unconnected particles."
1866	US astronomer Daniel Kirkwood noticed that a particle in the Cassini Division would be in 3:1 resonance with the period of Enceladus.

1872	Daniel Kirkwood was able to associate the Cassini Division and Encke's Division with resonances of the four then known interior moons: Mimas, Enceladus, Tethys and Dione.
1876	Spectacular white spots were observed on Saturn by US astronomer Asaph Hall.
1883	The first photograph of Saturn's rings was taken by British engineer and astronomer Andrew Ainslee Common.
1888	US astronomer James Edward Keeler became the first person to clearly observe the Encke Division (Encke only saw it as a dark band in 1837).
2 November 1889	Edward Emerson Barnard observed an eclipse of Iapetus by Saturn's rings; watching Iapetus become dimmer going through the shadow of the C-ring and disappear completely in the shadow of the B-ring, Barnard concluded that the C-ring must be semi-transparent and that the B-ring is opaque.
1895	James Keeler and William Campbell observed that the inner part of the rings orbit more rapidly than the outer part of the rings, confirming Maxwell's deduction of 1856 that Saturn's rings were made up of small particles.
1898	US astronomer William Henry Pickering discovered Phoebe, the first and only Saturn moon discovered from ground-based observations that were not made around the time of a ring-plane crossing.
15 June 1903	Edward Emerson Barnard observed a white spot on Saturn.
1911	Barnard photographed Saturn from Mt Wilson Observatory; the image shows that Saturn is visible through the A-ring.
1921	The first observation of a mutual satellite event, an eclipse of Rhea by Titan, was observed by L. Comrie and A. Levin.
1923	B. Lyot detected the polarization of light scattered by Saturn's rings.
1960	J. Botham observed a white spot on Saturn, and established a 30-year cycle for the white spot appearances.
1966	Stephen Larson, John Fountain and R. Walker discovered Epimetheus
	Audouin Dollfus discovered Janus; using the Pic du Midi Observatory, Dollfus was able to establish the actual edge-on brightness of the rings, and estimated the thickness of the rings as only 2.4 km.
1967	Walter Feibelman discovered the E-ring from Allegheny Observatory images taken the year before.

1969	Pierre Guerin found evidence of a possible D-ring.
1970	Measurements of the spectra of the rings in the near infrared strongly indicated the presence of water ice, which suggests the surfaces of the ring particles are predominantly water ice.
1977	Stephen James O'Meara observed dark radial features (which would later be called "spokes") on Saturn's rings and recorded them in a sketch. (Voyager 1 confirmed spokes in 1980.)
1978	H. Reitsema established the limits of the Encke Division with his observations of Iapetus as it was eclipsed by the rings.
1 September 1979	US space probe Pioneer 11 flew by Saturn 3500 km from the outer edge of the A-ring; it passed 20930 km from Saturn's cloud tops; discovered the F-ring; confirmed the E-ring.
1980	Brad Smith, S. Larson and R. Walker discovered Telesto.
1980	D. Pascu, P. Seidelmann, W. Baum and D. Currie discovered Calypso.
1980	P. Laques and J. Lecacheux discovered Helene.
1980	Bruno Sicardy and Andre Brahic were able to measure the thickness of the B-ring as about 1.1 km.
12 November 1980	US space probe Voyager 1 flew within 126000 km of Saturn's cloud tops; Voyager 1 was launched 5 September 1977.
1981	J. Lissauer and M. Henon proposed that moonlets are embedded in the ring system.
25 August 1981	US space probe Voyager 2 flew within 101000 km of Saturn's cloud tops, on its way toward Uranus and Neptune; Voyager 2 was launched 20 August 1977.
1990	While analyzing Saturn images taken by Voyager 2, Mark Showalter discovered a new moon, Pan, orbiting in the Encke Division of Saturn's rings.
1990	Stuart Wilbert observed the expected white spot on Saturn, previously seen in 1960.
22 May 1995	The first of a triple crossing of the Earth through the plane of Saturn's rings. The event lasted 24 minutes.
10 August 1995	The second of a triple crossing of the Earth through the plane of Saturn's rings.
11 February 1996	The last of a triple crossing of the Earth through the plane of Saturn's rings.
15 October 1997	At 8^h43^m UT, a seven-year journey to Saturn began with the liftoff of a Titan IVB/Centaur carrying the US–European–Italian Cassini space probe.
21 April 1998	The Cassini space probe flew by Venus for the first of two encounters.

25 June 2004	According to NASA's schedule, the Cassini space probe will undergo Saturn orbit insertion on this date. Then, on 27 November, the Huygens probe is scheduled to land on Titan.
4 September 2009	At 10^h19^m UT, the Earth will pass through the ring plane of Saturn. This is the first crossing since 1996. The heliocentric longitude of the Earth will be 342° and that of Saturn 174°.
23 March 2025	At 14^h38^m UT, the Earth will pass through the ring plane of Saturn. The heliocentric longitude of the Earth will be 183° and that of Saturn 351°.
15 October 2038	The first of three passages of the Earth through the ring plane of Saturn (a triple plane crossing event) will take place at 13^h41^m UT. The heliocentric longitude of the Earth will be 22° and that of Saturn 170°. The next will be at 23^h17^m on 1 April 2039 (heliocentric longitude of the Earth will be 191° and that of Saturn 175°), and the final one will be at 12^h43^m on 9 July 2039 (heliocentric longitude of the Earth will be 286° and that of Saturn 179°).

Uranus

Physical data

Size   51 118 km*

Mass 8.686×10^{25} kg

Escape velocity   21.30 km/s (76 680 km/hr)

Temperature range   Minimum   Average   Maximum
                          $-214\,°C$ $-205\,°C$ increases with depth

Oblateness   0.030

Surface gravity   7.77 m/s²

Volume 6.995×10^{13} km³
               64.4 times that of Earth

Magnetic field strength and orientation   Rapid rotation certainly aids the generation of a sizable magnetic field for Uranus. Unlike Jupiter and Saturn, both of whose metallic regions are near their respective cores, the metallic region of Uranus seems to be in the mantle of the planet. It has been suggested that this is the reason for the large offset of the magnetic field from the planet's core. The size of the magnetosphere of Uranus varies from 15–20 times the radius of the planet. Its composition is primarily hydrogen ions originating from the solar wind and from the atmosphere of Uranus.

Average surface field (tesla)	0.00003
Dipole moment (weber-meters)	3×10^{16}
Tilt from planetary axis	59°
Offset from planetary axis (planet radii)	0.3

Albedo (visual geometric albedo)   0.51

Density (water = 1)   1.29 g/cm³

Solar irradiance   3.8 watts/m²

Atmospheric pressure   varies with depth, $>10\,132\,500$ pascals

Composition of atmosphere

Molecular hydrogen (H_2)	$>82\%$
Helium (He)	$>14\%$
Methane (CH_3)	2%
Ammonia (NH_3)	0.01%
Ethane (C_2H_6)	0.00025%
Acetylene (C_2H_2)	0.00001%
Carbon monoxide (CO)	$<0.0000001\%$
Hydrogen sulfide (H_2S)	$<0.0000001\%$

Maximum wind speeds   720 km/hr

* Polar diameter 49 584 km.

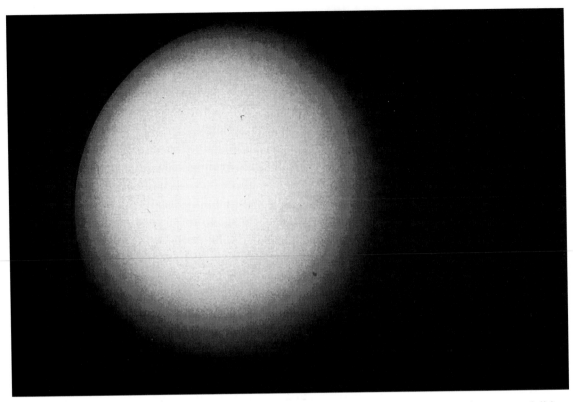

When Voyager 2 flew past Uranus in 1986, only a featureless haze was visible. (NASA)

Outstanding cloud features

The visible atmosphere of Uranus is generally a featureless haze. Throughout the history of our observations of the planet (now more than 217 years), a number of observers have seen detail on the planet. The first such observer, the British astronomer W. Buffham, in 1870, observed two round bright spots and a bright zone. The US astronomer Charles Augustus Young, in 1883, reported markings, along with both polar and equatorial belts. All detail was, of course, very faint.

In 1891, US astronomers Edward Singleton Holden and James E. Keeler and German astronomer John Martin Schaeberle, all saw bands during April 1891 using the Lick Observatory 91 cm refractor. In 1924, the French astronomer Eugène Marie Antoniadi saw grayish polar caps and two faint equatorial bands.

Observers making subsequent observations noticed the same sort of effects. Features were seen and then they would fade into invisibility. In 1970, images were taken with the balloon-riding Stratoscope II telescope. Even with a resolution of 0.1 second of arc, no detail was seen on Uranus.

The Voyager 2 mission to Uranus, in January, 1986, also showed a nearly featureless planet. The best recent detail-bearing image of Uranus was taken on 30 May 1993, at near-infrared wavelengths from the Multiple-Mirror Telescope near Mt Hopkins, AZ. The image showed a dark spot near the center of the disk of Uranus along with a bright region and a subtle, irregular dark band near the pole.

Orbital data

Period of rotation 0.71806 days
17.234 hours
$0^d17^h14^m$

Period of revolution (sidereal orbital period) 84.01 years
30684 days
$84^y03^d15.66^h$

Synodic period $369^d15^h50.4^m$

Equatorial velocity of rotation 8971.5 km/hr

Velocity of revolution 6.81 km/s (24516 km/hr)

Distance from Sun Average 19.1914 AU
2870990000 km
Maximum 20.105 AU
3007665000 km
Minimum 18.281 AU
2734799000 km

Distance from Earth Maximum 21.12 AU
3159769980 km
Minimum 17.26 AU
2582694020 km

Apparent size of Sun (average) 0.028°

Apparent brightness of Sun $m_{vis} = -20.3$

Orbital eccentricity 0.0461

Orbital inclination 0.774°

Inclination of equator to orbit 97.86°

Observational data

Maximum angular distance from Sun 180°

Brilliancy at opposition Maximum +5.65
Minimum +6.06

Angular size* Maximum 3.96″
Minimum 3.60″

* This measurement is the apparent angular equatorial diameter of Uranus, measured in seconds of arc, as seen from Earth. Note: Uranus's polar diameter is 97% of its equatorial diameter.

Sir William
Herschel. (Michael
E. Bakich
collection)

Discovery

For untold thousands of years, humanity knew of only five wanderers, or
planets, which traveled against the background of stars. But on 15 November
1738, Friedrich Wilhelm Herschel was born in Hanover, Germany. Forty-two
years and 118 days later – on 13 March 1781 – Herschel altered the celestial
landscape forever:

On Tuesday the 13th of March, between ten and eleven in the evening, while I was
examining the small stars in the neighbourhood of H Geminorum, I perceived one that
appeared visibly larger than the rest: being struck with its uncommon magnitude, I
compared it to H Geminorum and the small star in the quartile between Auriga and
Gemini, and finding it so much larger than either of them, suspected it to be a comet.

Illustration of the discovery of Titania and Oberon. Herschel, William, "An account of the discovery of two satellites revolving around the Georgian planet" *Philosophical Transactions*, **77**:125–9, London, 1787. (Linda Hall Library)

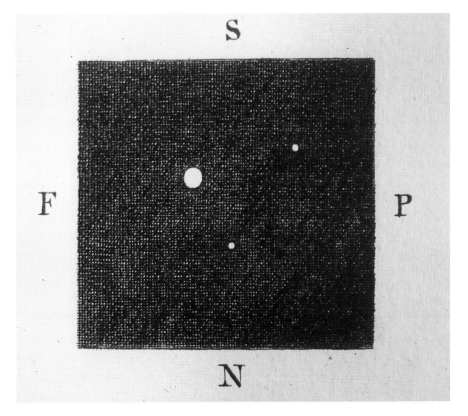

This is an excerpt from a paper read before the Royal Society in London on 26 April 1781 by Sir William Watson, Herschel's friend. Herschel had previously presented this on 28 March 1781 to the Bath Philosophical Society.

Herschel had initially assumed the new object to be a comet because it increased in apparent diameter as he increased the magnification of his telescope. His initial sighting was with an eyepiece which provided a magnification of 227. To prove that the object was not a star, Herschel first increased the magnification to 460, and then to 932. In Herschel's day, new comets were sighted infrequently, but they were not once-in-a-lifetime events.

Herschel's proof that the object was a "comet" came during his next observing session, on 17 March 1781. He checked the position of the object and saw that it had indeed moved. The next day, Herschel communicated his discovery to Nevil Maskelyne, the Astronomer Royal of England. Maskelyne made a number of observations of the object, and was unconvinced of its cometary nature. In fact, on 4 April he wrote to Watson and referred to the object as a "comet or new planet." On 23 April, he used the same terminology in a letter to Herschel. Maskelyne had been unable to see any fuzziness in the appearance of the object, and he certainly had seen no tail, a sighting which would have conferred the status of comet.

The German astronomer Johann Elert Bode began a series of observations of Herschel's object, and by November 1781, he was convinced that Herschel

Herschel's 40-foot
reflector. (Michael
E. Bakich
collection)

had, indeed, discovered a new planet. Bode even worked out an orbit for
"Planet Herschel." He found, amazingly, that it lay twice as far from the Sun
as Saturn. Bode was the first to suggest the name Uranus. His reasoning was
mythological – Saturn was the father of Jupiter, so the next planet out should
be the father of Saturn.

Herschel had not discovered a comet, but rather the first new planet since
antiquity. The astronomical community was delighted. With this one observation, Herschel had doubled the size of the solar system. Also, here was
another planet which could be used to test the relatively new law of universal gravitation.

The honor of naming the new planet naturally fell to Herschel. (This was
more than a century before any thought was given to planetary nomenclature committees.) Herschel proposed the name Georgium Sidus (George's
Star) in honor of George III, then King of England. In fact George III, when
made aware of Herschel's discovery, granted Herschel an annual stipend.

266

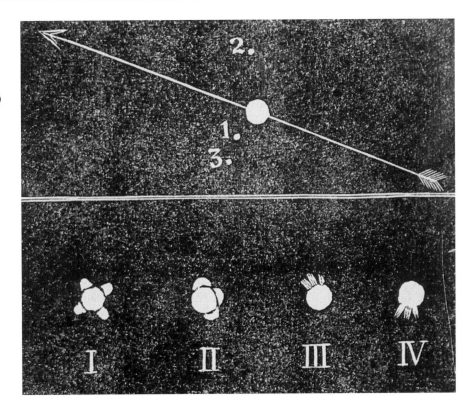

Uranus, showing rings! Breen, James, *The Planetary Worlds*, London, 1854. (Linda Hall Library)

George's only requirement was that Herschel move to a location not too distant from Windsor. This being accomplished, George reasoned, members of the royal family would have access to the wonderful views afforded by Herschel's telescopes.

As may be imagined, outside of England the thought of a major member of the solar system named after a British monarch was met with less-than-enthusiastic support. Many non-British astronomers took to calling the new planet "Herschel." Some years after Herschel's death (in 1822), Bode's suggestion of Uranus as the name for the new world finally gained the total support of the astronomical community.

Important concepts

The rings and their discovery from Earth

In 1973, British astronomer Gordon Taylor of the Royal Greenwich Observatory predicted that, on 10 March 1977, Uranus would occult SAO 158687, a 9th magnitude star located in the constellation Libra. Taylor provided locations where this event would be visible and two groups of astronomers decided that the opportunity to learn more about Uranus was too good to let pass.

An event such as this, observed photoelectrically, would yield data as the star's light passed through the upper atmosphere of Uranus. Properties of the atmosphere, and how these properties changed with depth, could be analyzed. In addition, physical characteristics of the planet, such as its precise size, could be studied.

Located at the Perth Observatory in Australia were Bob Millis, Peter Birch and Dan Trout, all from Lowell Observatory. A second group, composed of James L. Elliot, Edward Dunham and Doug Mink, all of Cornell University, was aboard the Kuiper Airborne Observatory, administered by the United States' National Aeronautics and Space Administration (NASA).

The Kuiper Airborne Observatory crew recorded continuous occultation data for 2 hours 12 minutes. Just prior to the immersion of SAO 158687, several spikes were seen in the photometric plot. Similar light curve features had been seen during occultations of Jupiter and Neptune, and in those cases thin planetary rings were the cause. Could the same be true for Uranus? Eagerly, the crew waited for the emersion of the star, and, indeed, several more spikes were seen.

A message was quickly sent to Perth Observatory to urge the Lowell astronomers to extend their photometric observations beyond the re-appearance of the star from behind the disk of Uranus. Later, when the data was examined, it was apparent that there were five dips in the star's light before its occultation by Uranus, and that these were exactly matched by five downward spikes after the occultation. Thanks to an early prediction which allowed careful preparations to be made, rings around Uranus had been discovered and confirmed.

Why is Uranus's axis tilted so far?

Uranus is distinguished by the fact that its axis is tilted by 97.86°. Pluto's axis is tilted by an additional 24.66°, but the obliquity of Pluto has been attributed by a number of astronomers to influences from its satellite, Charon. Charon possesses approximately 10% the mass of Pluto, a higher satellite-to-planet mass ratio (1:7.6) than is found elsewhere in the solar system. The nearest approach to this value is the Moon–Earth ratio, which is 1:82.5.

Today, current thought is that Uranus did not form with such a high obliquity. Planetary scientists believe that at some point during its formation, a planet-sized body collided with Uranus, giving it the currently observed axial tilt. Calculations show that the impactor probably had a mass equal to that of the Earth, and that it struck Uranus near one of its poles. If the impactor had been composed of ice, it would have added significantly to the water content of the Uranian atmosphere, and may be the reason for the lack of an internal heat source within the planet.

Voyager 2 found that one of the most striking influences of the Uranian obliquity is its effect on the tail of the planet's magnetic field, which is itself tilted 60° from the axis of rotation. The magnetotail was shown to be twisted into a long corkscrew shape behind the planet. The source of the magnetic field, the electrically conductive, super-pressurized ocean of water and ammonia once thought to lie between the core and the atmosphere has not been confirmed.

Interesting facts

Uranus was discovered by the German-born British astronomer Sir William Herschel, on 13 March 1781. Herschel at first believed he had discovered a comet.

The Sun is 363 times as bright from the Earth as from Uranus.

The orbit of Uranus is tilted less than 1° from the plane of the ecliptic.

Uranus's average apparent motion (against the background of stars) is approximately 42 seconds of arc per day. So it takes Uranus about 44 days to move the width of the Full Moon.

Uranus is 14½ times as massive as the Earth.

The dayside temperature of Miranda reaches −187°C.

If Uranus were hollow, it could hold 64 Earths.

Observing data

Future dates of conjunction

6 Feb 2000
9 Feb 2001
13 Feb 2002
17 Feb 2003
22 Feb 2004
25 Feb 2005
1 Mar 2006
5 Mar 2007
9 Mar 2008
13 Mar 2009
17 Mar 2010

Future dates of opposition

7 Aug 1999
11 Aug 2000
15 Aug 2001
20 Aug 2002
24 Aug 2003
27 Aug 2004
1 Sep 2005
5 Sep 2006
9 Sep 2007
13 Sep 2008
17 Sep 2009
21 Sep 2010

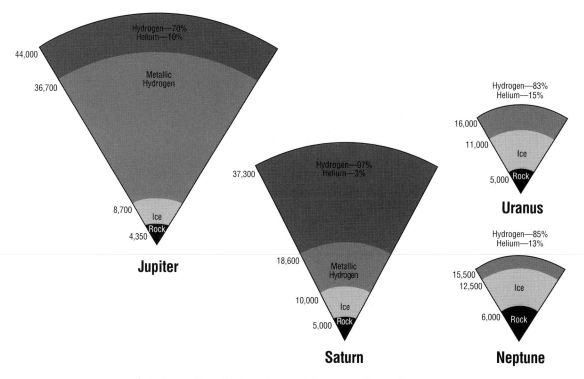

Interiors of the Jovian planets (distances from planet centers are in kilometers). (Holley Bakich)

Recent data

The Voyager 2 spacecraft was launched by the United States in 1977. Its encounter with Uranus began 4 November 1985 with an observing phase. Activity built to a peak in late January 1986, with most of the critical observations occurring in a six-hour period in and around the time of closest approach. The spacecraft made its closest approach to Uranus at 17:59 UT on January 24. At that moment it was within 81 500 km of the planet's cloud tops.

As expected, the dominant constituents of the atmosphere are hydrogen and helium. But Voyager 2 found that only 15 percent of the atmosphere was helium. This percentage was considerably less than the 40 percent that had been suggested by some Earth-based studies. Methane, acetylene and other hydrocarbons exist in much smaller quantities. Methane in the upper atmosphere absorbs light with longer (reddish) wavelengths, giving Uranus its blue-green color.

The atmosphere of Uranus is arranged into clouds running at constant latitudes, similar to the more vivid bands seen on Jupiter and Saturn. Winds at mid-latitudes on Uranus blow in the same direction as the planet rotates, just as on Earth, Jupiter and Saturn. Velocities are in the range of 40–160 m/s. At the equator, winds of about 100 m/s were found, blowing in the opposite direction.

A high layer of haze, essentially a photochemical smog, was detected

The surface of
Miranda. (NASA)

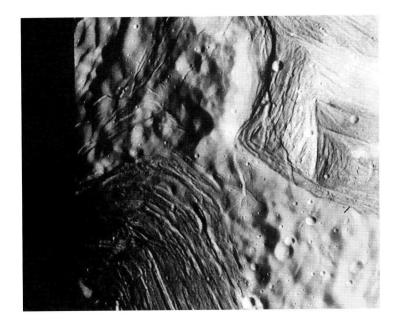

around the sunlit pole. The sunlit hemisphere also was found to radiate large amounts of ultraviolet light, a phenomenon that Voyager scientists have dubbed "dayglow."

The average temperature on Uranus is about 68 K. The minimum near the tropopause is 59 K where the atmospheric pressure is 10133 pascals. Surprisingly, both the sunlit and dark poles and, in fact, most of the planet, show nearly the same temperature below the tropopause. The temperature rises with increasing altitude, reaching 150 K in the rarefied upper atmosphere.

Voyager 2 found strong radiation belts circling Uranus dominated by hydrogen ions. Uranus's radiation belts are so intense that irradiation from them would darken any methane trapped in the icy surfaces of the inner moons and ring particles. Such darkening would occur rather rapidly, in planetary terms. The full effect would be seen after "only" 100000 years. This phenomenon may have contributed to the dark appearances of those objects.

Prior to the flyby, the rotational rate of Uranus was known to an accuracy of approximately 5 percent. Voyager 2 detected radio emissions from Uranus that, along with imaging data, helped refine the planet's rate of rotation to 17 hours 14 minutes.

Voyager 2 obtained clear, high-resolution images of each of the five large satellites of Uranus known before the encounter (Miranda, Ariel, Umbriel, Titania and Oberon) and it discovered 10 additional moons. The largest of the newly detected moons, named Puck, is 154 km in diameter.

Preliminary analysis shows that the large moons are ice–rock conglomerates like the satellites of Saturn. These satellites are made up of about 50 percent water ice, 20 percent carbon- and nitrogen-based materials, and 30 percent rock. Their surfaces, almost uniformly dark gray in color, display varying degrees of geologic history. Very ancient, heavily cratered surfaces are apparent on some of the moons, while others show strong evidence of internal geological activity.

Miranda, innermost of the five large moons, is one of the strangest bodies in the solar system. Its surface consists of huge fault canyons as deep as 20 km, terraced layers and a mixture of old and young surfaces. The younger regions may have been produced by incomplete differentiation of the moon, a process in which upwelling, lighter material surfaced in limited areas. Alternatively, Miranda may be a re-aggregation of material from an earlier time when the moon was fractured into pieces by a violent impact. Given Miranda's small size and low temperature, it is believed that an additional energy source such as tidal heating caused by the gravitational tug of Uranus must have been involved.

In 1977, the first nine rings of Uranus were discovered. During the Voyager encounters, those rings were photographed and measured, as were two other new rings. These were designated 1986U1R, which was detected between the two outermost of the previously known rings, and 1986U2R, a broad region of material 3000 km across.

Rings of Uranus

Name	Distance (km)[a]	Width (km)
1986U2R	38 000	2500
6	41 840	1–3
5	42 230	2–3
4	42 580	2–3
Alpha	44 720	7–12
Beta	45 670	7–12
Eta	47 190	0–2
Gamma	47 630	1–4
Delta	48 290	3–9
1986U1R	50 020	1–2
Epsilon	51 140	20–100

Note:

[a] The distance is measured from the center of Uranus to the start of the ring.

The thickness of each ring, with the exception of Epsilon, is 0.1 km. The Epsilon ring is slightly thicker, at 0.15 km. A very tenuous distribution of fine dust also seems to be spread throughout the ring system.

Incomplete rings and the varying opacity in several of the main rings lead scientists to believe that the ring system may be relatively young and did not form at the same time as Uranus. The particles that make up the rings may be remnants of a moon that was broken by a high-velocity impact or torn up by tidal effects.

The Uranian ring system. (NASA)

The most interesting of these rings, the Epsilon ring, is composed mostly of ice boulders up to a meter across. This ring is surprisingly deficient in particles smaller than about the size of a beach ball. This may be due to atmospheric drag from the planet's extended hydrogen atmosphere, which probably siphons smaller particles and dust from the ring.

Important clues to Uranus's ring structure may come from the discovery that two small moons – Cordelia and Ophelia – straddle the Epsilon ring. This finding lends credence to theories that small moonlets may be responsible for confining or deflecting material into rings and keeping it from escaping into space. Eighteen such satellites (two for each of the previously known rings) were expected to have been found, but only two were photographed.

In 1997 the discovery of two additional Uranian satellites by B. J. Gladman, P. D. Nicholson, J. A. Burns and J. J. Kavelaars was announced. They have been given the designations S/1997 U1 and S/1997 U2. Both were detected on CCD frames taken 6–7 September 1997. Calculations in December, 1997, indicated that both satellites have retrograde orbits about Uranus at a distance of approximately 5740000 km with periods of about 415 days. For an assumed albedo of 0.07, the sizes of these new bodies would be about 40 and 80 km. The discovery of these satellites places Uranus more firmly in the realm of the Jovian planets, as, prior to this find, it was the only one not known to possess distant satellites. At the time of writing, these moons have not been named.

Satellites

	Size (km)	Mass (kg)[a]	Density (g/cm³)	Orbital period	Eccentricity	Inclination	Distance from planet (km)
Cordelia	26	—	—	08^{h}02.4^m	0.0005	0.14°	4.9752 × 10⁴
Ophelia	32	—	—	09^{h}02^m	0.0101	0.091°	5.3764 × 10⁴
Bianca	44	—	—	10^{h}25.8^m	0.0009	0.156°	5.9165 × 10⁴
Cressida	66	—	—	11^{h}07.5^m	0.0001	0.282°	6.1777 × 10⁴
Desdemona	58	—	—	11^{h}22.1^m	0.0002	0.16°	6.2659 × 10⁴
Juliet	42	—	—	11^{h}50^m	0.0002	0.042°	6.4358 × 10⁴
Portia	110	—	—	12^{h}19^m	0.0002	0.087°	6.6097 × 10⁴
Rosalind	58	—	—	13^{h}24.2^m	0.0006	0.057°	6.9927 × 10⁴
Belinda	68	—	—	14^{h}57.9^m	0.0001	0.033°	7.5255 × 10⁴
Puck	154	—	—	18^{h}17^m	0.0001	0.314°	8.6004 × 10⁴
Miranda	480(× 468 × 465)	6.59 × 10¹⁹	1.15	01^{d}09^{h}54.7^m	0.0027	4.22°	1.298 × 10⁵
Ariel	1162(× 1156 × 1156)	1.353 × 10²¹	1.56	02^{d}12^{h}28.8^m	0.0034	0.31°	1.298 × 10⁵
Umbriel	1169	1.172 × 10²¹	1.52	04^{d}03^{h}27.4^m	0.0050	0.36°	2.66 × 10⁵
Titania	1578	3.527 × 10²¹	1.70	08^{d}16^{h}56.6^m	0.0022	0.10°	4.358 × 10⁵
Oberon	1523	3.014 × 10²¹	1.64	13^{d}11^{h}06.7^m	0.0008	0.10°	5.826 × 10⁵

Note:

[a] Satellites with no value for their mass have never had this quantity measured to any degree of accuracy.

Historical timeline	13 March 1781	Hanover-born British astronomer Sir William Herschel discovered Uranus within the boundaries of the constellation Gemini.
	11 January 1787	Herschel discovered Titania and Oberon.
	1821	French astronomer Alexis Bouvard studied the orbit of Uranus; he found a discrepancy between predicted and observed positions.
	1828	Observed positions of Uranus were so far off predictions that astronomers began a search for a trans-Uranian planet.
	1843	John Couch Adams calculated the position of Neptune from observed irregularities in the orbit of Uranus.
	24 October 1851	British astronomer William Lassell discovered Ariel and Umbriel.
	20 December 1852	While observing Uranus, Lassell suspected a spot and an equatorial dark band, but was too good an observer to state that he had definitely seen them.
	25 January 1870	British astronomer W. Buffham became the first observer to record definite markings on Uranus.
	16 February 1948	Dutch-born US astronomer Gerard Peter Kuiper discovered Miranda.

10 March 1977	James L. Elliot, Edward Dunham and Doug Mink, flying aboard the Kuiper Airborne Observatory, discovered the rings of Uranus when the planet occulted the 9th magnitude star SAO 158687, which lies within the boundaries of the constellation Libra.
30 December 1985	US astronomer Stephen Synnott discovered Puck.
3 January 1986	Voyager 2 scientists discovered Portia and Rosalind.
9 January 1986	Voyager 2 scientists discovered Juliet.
13 January 1986	Voyager 2 scientists discovered Belinda, Cressida and Desdemona.
20 January 1986	Voyager 2 scientists discovered Cordelia and Ophelia.
21 January 1986	Voyager 2 scientists discovered Bianca.
24 January 1986	Voyager 2 flew within 81 433 km of the cloud tops of Uranus, on its way to Neptune; Voyager 2 was launched 20 August 1977.
October 1986	John Clarke explained the mechanism causing what is known as dayglow in the atmosphere of Uranus.
6–7 September 1997	B. J. Gladman, P. D. Nicholson, J. A. Burns and J. J. Kavelaars announced the discovery of Uranian satellites S/1997 U1 and S/1997 U2.

Neptune

Physical data

Size   49 572 km*

Mass 1.024×10^{26} kg

Escape velocity   23.50 km/s (84 600 km/hr)

Temperature range   Minimum   Average   Maximum

$-223\,°C$ $-220\,°C$ increases with depth

Oblateness   0.026

Surface gravity   11.0 m/s^2

Volume 6.379×10^{13} km^3
58.7 times that of Earth

Magnetic field strength and orientation   Neptune's magnetic field is almost an exact duplicate of that of Uranus in size, strength, inclination to planetary axis and offset from the center of the planet. The magnetosphere, like that of Uranus, is composed of hydrogen ions that originate from Neptune's atmosphere and from the solar wind.

Average surface field (tesla)	0.000 02
Dipole moment (weber-meters)	2×10^{16}
Tilt from planetary axis	55°
Offset from planetary axis (planet radii)	0.5

Albedo (visual geometric albedo)   0.41

Density (water = 1)   1.64 g/cm^3

Solar irradiance   1.5 watts/m^2

Atmospheric pressure   varies with depth, > 10 132 500 pascals

Composition of atmosphere

Molecular hydrogen (H_2)	> 84%
Helium (He)	> 12%
Methane (CH_3)	2%
Ammonia (NH_3)	0.01%
Ethane (C_2H_6)	0.000 25%
Acetylene (C_2H_2)	0.000 01%
Carbon monoxide (CO)	< 0.000 000 01%
Hydrogen sulfide (H_2S)	< 0.000 000 01%

Maximum wind speeds   2400 km/hr

* Polar diameter 48 283 km.

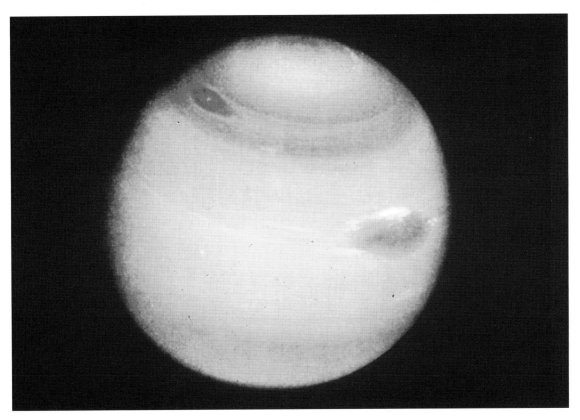

Neptune from Voyager 2. (NASA)

Outstanding cloud features

The Great Dark Spot is certainly the most noticeable feature within the atmosphere of Neptune. It is an anticyclone similar to Jupiter's Great Red Spot. Neptune's Great Dark Spot may be found in the above photograph at latitude 22° south, also the same as the Great Red Spot.

At about 42° south, a bright, irregularly shaped, eastward-moving cloud circles much faster than does the Great Dark Spot, "scooting" around Neptune in about 16 hours. This "scooter" may be a cloud plume rising between cloud decks.

Another spot, designated "DS2" by Voyager's scientists, is located far to the south of the Great Dark Spot, at 55° south. It is almond-shaped, with a bright central core, and moves eastward around the planet, also in about 16 hours.

Long, bright clouds, reminiscent of cirrus clouds on Earth, can be seen high in Neptune's atmosphere. They appear to form above most of the methane, and consequently are not blue.

At northern low latitudes (27° north), the Voyager 2 spacecraft captured images of cloud streaks casting their shadows on cloud decks estimated to be about 50–100 km further down. The widths of these cloud streaks range from 50–200 km, and the widths of the shadows range from 30–50 km. Cloud streaks were also seen in the southern polar regions (71° south), where the cloud heights were about 50 km.

Clouds in the
atmosphere of
Neptune. (NASA)

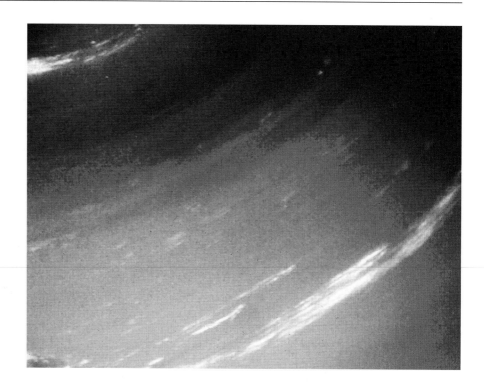

Orbital data

Period of rotation	0.671 25 days
	16.11 hours
	$0^d16^h6.5^m$

Period of revolution (sidereal orbital period) 164.79 years
60 188.3 days
$164^y288^d13^h$

Synodic period $367^d11^h45.6^m$

Equatorial velocity of rotation 9667.1 km/hr

Velocity of revolution 5.45 km/s (19 620 km/hr)

Distance from Sun	Average	30.0611 AU
		4 504 300 000 km
	Maximum	30.3106 AU
		4 534 406 000 km
	Minimum	29.805 AU
		4 458 765 000 km

Distance from Earth	Maximum	31.33 AU
		4 686 510 980 km
	Minimum	28.79 AU
		4 306 660 020 km

Apparent size of Sun (average)   0.017°

Apparent brightness of Sun $m_{vis} = -19.1$

Orbital eccentricity   0.0097

Orbital inclination   1.774°

Inclination of equator to orbit   28.31°

Observational data

Maximum angular distance from Sun   180°

Brilliancy at opposition   Maximum   +7.66
          Minimum   +7.70

Angular size*   Maximum   2.52″
          Minimum   2.49″

Discovery

The most amazing things can happen when one begins to compile data. Take the case of the French astronomer Alexis Bouvard. Bouvard was an assistant to Pierre Simon de Laplace and even helped with calculations for Laplace's epic *Traité de la Méchanique Céleste*. In 1820, Bouvard undertook the task of compiling positional tables for Uranus. He set out to define the entire orbit of that planet from observations made during the nearly forty years that had elapsed since its discovery in 1781.

Bouvard found that Uranus had been observed at least fifteen times before its discovery by Herschel. Indeed, between 1690 and 1771, it had been recorded in catalogs as a fixed star. Knowing that earlier positions would help to refine the orbit, Bouvard welcomed the fact that the planet had been so often seen before. On the completion of his calculation, however, he met with a most extraordinary and puzzling circumstance. The orbit he deduced was found unsatisfactory to either the pre-discovery or the post-discovery set of observations, and its deviation from the older ones was striking.

Bouvard, quite understandably, assumed that one of the sets of observations must suffer from a lack of accuracy. He thought it most logical that it would be the earlier set:

Such is the alternative presented by the formation of the tables of the planet Uranus, that if the ancient observations are compared with the modern ones, the first are passably represented, while the second are not so with the precision they demand; and if either set be rejected, tables are the result which satisfy the ones retained, but do not satisfy the others. It being then necessary to decide between them, I have held by the modern observations, as being the most likely to be accurate, and I leave Time to come in aid of the difficulty of reconciling these older ones, and of explaining whether it is caused by the inexactness of these old observations, or depends on some foreign and yet unperceived influence to which the planet is subjected. [J. P. Nichol, *The Planet Neptune*, 1846]

* This measurement is the apparent angular diameter of Neptune, measured in seconds of arc, as seen from Earth.

This conclusion was a bold one, and Bouvard did not arrive at it lightly. He knew the names and reputations of some of the observers whose precision he was questioning – names like Mayer, Flamsteed and Bradley. This was certainly not the solid ground Bouvard had hoped for on which to explain his results. There had to be another answer, and nineteenth-century astronomers were quite willing to propose – some might say *guess* – what that answer would be.

One of the first explanations to be suggested was that the discrepancies resulted from disorder, superimposed upon the order of the solar system. What was the form of this disorder? A comet! One of these wandering bodies, it was said, had, in the course of its devious path through our system, come into contact with Uranus – struck it, in fact. Thus, by introducing a new cause of motion the comet produced the discrepancy discerned between the orbit of the planet at the epoch of the older observations and its movements during Bouvard's time.

Unfortunately, Bouvard found that there was not simply one deviation in the orbit of Uranus, but a continuous series of them. Astronomers might be persuaded to imagine one impact and its effect. However, to expect them to believe more than one such cataclysmic event is to ask that the laws of probability be completely overturned.

Soon after Bouvard's results were made public, a number of astronomers wondered if too much faith had been placed in Newton's law of universal gravitation. Perhaps "universal" was a bit too broad. Certainly it seemed that gravitation behaved as expected out to the limit of Saturn, but the discovery of Uranus had pushed the bounds of the solar system to new limits. Could it be that gravity behaved differently at such a great distance? This so offended the sensibilities of most astronomers that it was determined that every other avenue would be exhausted before a new law of gravitation was proposed.

Another approach was to revise, in part, the theory given by Laplace concerning the gravitational influence of Jupiter and Saturn. This was done by substituting different values for the masses, etc. of these planets. Those results were then applied to Bouvard's original calculations. It must have been a tiring procedure, and it was to no avail. The irregularities in the motion of Uranus remained.

Just as some astronomers were giving up, however, others were entering the fray. In the midst of the calculation frenzy was a respected astronomer who finally hit upon the reason for the "error." His name was LeVerrier.

Urbain Jean Joseph LeVerrier was a well-respected astronomer who worked at the Paris Observatory. He had published his first paper in 1832, a treatise on meteors. He had attempted to reconcile irregularities in Mercury's orbit by proposing, in 1860, the existence of an intra-Mercurial planet which he eventually named Vulcan. Now, because of a suggestion by Paris Observatory Director Dominique François Arago, his mind was turned to the orbit of Uranus.

LeVerrier quickly reduced the problem to a simpler form: "Can the anomaly be explained by the supposed action of a foreign and hitherto unknown body on Uranus?" Of course, that naturally leads to a follow-up question: "Is the

Urbain Jean
Joseph LeVerrier.
(Michael E. Bakich
collection)

foreign body a new planet, or is it a body connected with Uranus – a satellite?"
The idea of the perturbing influence being a moon found favor with some
astronomers not familiar with the intricacies of celestial mechanics. However,
if Uranus was so disturbed by a satellite, that satellite must be large and would
have been observed even with the telescopes of the time. Of more importance,
however, is the fact that the type of perturbations that Uranus was undergoing could not have been produced by a satellite.

After a lengthy series of calculations, LeVerrier wrote a paper* in which his
first conclusion was as follows:

There is in the whole ecliptic only one region in which the perturbing planet can be
supposed to be placed, so that it accounts for the irregular movements of Uranus. On
the First of January, 1800, its mean longitude must have been between 243° and 252°.

So in his first estimate, LeVerrier narrowed the search to a scant 9° of the ecliptic. This was not good enough, however.

* Quoted in J. P. Nichol, *The Planet Neptune*, 1846.

His next published assertion*, involving many more calculations, was more exact:

That all the observed motions of Uranus could be accounted for by the perturbing action of a planet, the elements of whose orbit more primarily assumed, whose longitude on 1st January 1800 is 252°, and whose eccentricity and the longitude of its perihelion were determined by processes he had just explained.

LeVerrier had assumed a mass for the new planet between one and three and a half times as great as that of Uranus. From this, it followed that on 1 January 1847, the heliocentric longitude of the unknown planet must be 325°.

On 31 August 1846, LeVerrier produced his third paper on the subject of a trans-Uranian planet. He included more early observations and refined the position of Uranus for 1 January 1847 to 326°32', instead of 325°. A contemporary account* of LeVerrier delivering this paper before the French Academy of Sciences is particularly poignant:

A young man, not yet at life's prime, speaking unfalteringly of the necessities of the most august Forms of Creation – passing onwards where Eye never was, and placing his finger on that precise point of Space in which a grand Orb lay concealed; having been led to its lurking-place by his appreciation of those vast harmonies, which stamp the Universe with a consummate perfection! Never was there accomplished a nobler work, and never work more nobly done! It is the eminent characteristic of these labours of LeVerrier, that at no moment did his faith ever waver: the majesty of the Enterprise was equaled by the resolution and confidence of the Man. He trod those dark spaces as Columbus bore himself amid the waste Ocean; even when there was no speck or shadow of aught substantial around the wide Horizon – holding by his conviction in those grand verities, which are not the less real because above sense, and pushing onwards towards his New World!

LeVerrier communicated the position of the object to a number of observatories around the world, but it was the Berlin Observatory which was particularly equipped to make the discovery. The astronomers there, led by C. Bremiker, were producing a new, detailed star chart centered on the ecliptic. The region of sky where LeVerrier had asked for a search to be conducted had just been completed and was with the printer. The observatory obtained a proof sheet and Johann Gottfried Galle discovered the planet on the very evening of the day on which he received the letter from LeVerrier indicating its place. The following is his reply to LeVerrier:

Berlin, 25th September, 1846

Sir,

The Planet whose position you marked out actually exists. On the day on which your letter reached me, I found a star of the eighth magnitude, which was not recorded in the excellent map designed by Dr. Bremiker, containing the twenty-first hour of the collection published by the Royal Academy of Berlin. The observations of the succeeding day showed it to be the Planet of which we were in quest . . .

J. G. Galle

* Quoted in J. P. Nichol, *The Planet Neptune*, 1846.

John Couch
Adams. (Michael E.
Bakich collection)

The heliocentric longitude of the new planet, as ascertained by Galle, was 327°24', for the epoch of 1 January 1847. LeVerrier had predicted a position of 326°32'. The difference, therefore, was less than 1°! Truly this was a triumph for LeVerrier and celestial mechanics, for Newton and the law of universal gravitation, for Galle and the Berlin Observatory, and for a modest man from Laneast, Cornwall, England, named John Couch Adams.

Adams had demonstrated an adeptness in astronomy at an early age and few of his contemporaries were his peer in mathematics. After graduating from Cambridge University in 1843 – with top honors – Adams began a numerical search for the planet responsible for the irregularities in the orbit of Uranus. He completed the bulk of his research by October 1843, and first worked out an approximate position for the trans-Uranian body in the middle of 1845.

In a letter dated 22 September 1845, James Challis, Professor of Astronomy at Cambridge University, wrote to the Astronomer Royal, Sir George Biddell Airy regarding the calculations of his former student, Adams. In the letter, he mentions that Adams would call on Airy personally. Adams did so on 22 September, but Airy was in France. He returned on 26 September. Adams's second call was on 21 October. He actually visited the Airy residence twice. The first time Airy was not at home, and the second time the butler refused him admittance because the Airy family was at dinner. Adams left a letter in which he provided Airy the results of his work.

In a letter dated 5 November 1845, Airy wrote to Adams in Cambridge, requesting more information about the radius vector of the hypothetical planet. Adams never replied to Airy's letter, nor supplied the requested information. Adams later confided to J. W. L. Galisher that he would have replied to Airy's request, but he considered the request unimportant to the overall problem. The next correspondence between Adams and Airy was in September 1846, when Adams provided a detailed account of his work. It was then a week after Galle had found LeVerrier's planet.

A firestorm ensued. Airy was berated publicly, privately, professionally, . . . are there any other means? Rest assured, Airy was assailed in those ways as well. The correspondence among astronomers during the latter part of 1846 is among the most heated in the whole history of the science. Airy's treatment, in some ways, rivals that of Galileo. And, as passionate as the arguments in England, Airy himself nearly turned it into an international affair only three weeks after the discovery by writing to LeVerrier and (very politely) challenging his claim to priority in the discovery.

Adams endured the initial onslaught. He also had some powerful supporters, most notable among them, Sir John Herschel. As the furor subsided and the truth came out, both LeVerrier and Adams were honored for the most amazing theoretical discovery in the history of astronomy. Later in life Adams, by all accounts a humble and unassuming man, refused a knighthood offered by Queen Victoria and also declined to be named Astronomer Royal, citing old age as the reason.

The following is an excerpt from Allan Chapman, "Private research and public duty: G. B. Airy and the search for Neptune" *Journal for the History of Astronomy*, **19** (1988) 121–39. It shows a side to the story that many have not considered.

Popular interpretations of this incident place a great deal of responsibility upon Airy, for not having taken the initiative to secure a British discovery. Yet this is unjust, and several key factors must be born in mind:

1. It was not the job of the Astronomer Royal to undertake searches.
2. As an extremely over-worked man, Airy cannot be blamed for being unavailable when Adams chanced to call upon him without first having made an appointment. He was abroad on the first occasion, and at dinner with his family on the second.
3. Airy's letter of 5 November 1845 questions Adams about the radius vector. What purpose was Airy's request? A letter from Airy to James Challis in December,

George Biddell
Airy. (Michael E.
Bakich collection)

1846, states, "If no adequate radius vector value had been available then the theory would have been false, not from any error of Adams's but from a failure in the law of gravitation. On this question therefore turned the continuance or fall of the law of gravitation."

4. Why was Adams not admitted when Airy was at dinner? We should bear in mind that at the time Mrs Richarda Airy was within a week of giving birth to their ninth child. Her previous pregnancies had been difficult, and as Airy was deeply attached to his wife, he saw no reason to have their dinner interrupted by a stranger who wished to see him on a business matter. There is no evidence to suggest that Adams was willing to wait until the meal was over in spite of the fact that the Airy family dined not in the evening, but in the late afternoon.

5. Airy's voluminous surviving correspondence makes it clear that everyone – from Cabinet Ministers and Admirals, down to servant-girls wanting to have their fortunes told – wrote to, and occasionally called-in upon the Astronomer Royal. A man who was so much in the public eye had to defend his privacy.

6. While all of this was going on, the Royal Observatory was being rocked by the disclosure of an awful incident. Airy's journal for 27 October 1845 states, "Investigated a very serious charge of incest against Mr Richardson, and suspended him from office." William Richardson was a senior Greenwich Observatory Assistant. Airy and his Chief Assistant, Robert Main, made appearances before the courts at the beginning of Richardson's trial. From Airy's diary, 24 February 1846: "Mr Main absent today at the Old Bailey before the Grand Jury, on the trial of W. Richardson for the willful murder of his incest child." Airy must have been embarrassed by the regular appearance of his name in the newspaper columns during the scandal.

7. 1845–1846 was probably the busiest year in Airy's professional life. As the Scientific Commissioner of the Railway Gauge Commission, he was travelling around Britain testing trains and interviewing engineers. It was this Commission which settled British (and American) railway gauges at the "Standard Gauge" of 4 feet 8½ inches.

8. Urbain LeVerrier had the determination to see his computations put to effect. Yet even he was not able to find a French Observatory that was willing to undertake the search, and was forced to write to colleagues in Berlin. We often forget that the French scientific establishment let LeVerrier down no less than the British was accused of having let down Adams. Once the Berlin sighting had been made, however, the French were quick to turn it into a French National discovery.

9. Airy never doubted LeVerrier's real primacy in the discovery. Though Adams had got his results first, credit went to LeVerrier because he published and acted on his. In Airy's mind, scientific knowledge was, by definition, public, open, international and broadly useful. The prize, therefore, went not to the man who hoarded his discoveries, but to him who followed them through to public announcement and the advancement of learning.

10. One of the reasons why Airy's conduct touched upon raw nerves, especially in the autumn of 1846, was his openhanded willingness to back a Frenchman's claim while seemingly doing nothing to back that of an Englishman. Airy's seeming lack of patriotic commitment did not sit well with his countrymen.

Historians will continue to debate the conduct of Airy, Adams and everyone else associated with this story. It is left to the gentle reader to decide who was at fault, or indeed, if there was any fault at all.

Important concepts

The Great Dark Spot

Even though Neptune receives only 3 percent as much sunlight as Jupiter, it is still a dynamic planet. Several large, dark spots imaged by Voyager 2 in 1989 were reminiscent of Jupiter's hurricane-like storms. The largest spot was big enough for Earth to fit neatly inside it. Named the Great Dark Spot by its discoverers, the feature appeared to be an anticyclone similar to Jupiter's Great Red Spot.

Neptune's Great Dark Spot was comparable in size, relative to the planet,

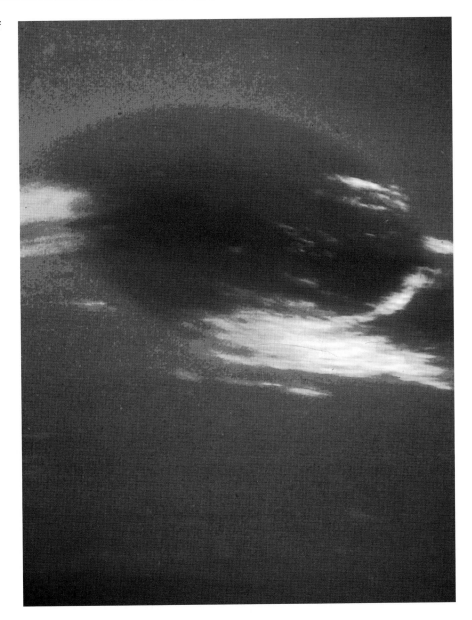

Voyager 2 image of Neptune's Great Dark Spot which disappeared in 1994. The spot is approximately 15 000 km across. (NASA)

and at the same latitude (the Great Dark Spot was at 22° south latitude) as Jupiter's Great Red Spot. However, Neptune's Great Dark Spot was more variable in size and shape than its Jovian counterpart. Bright, wispy clouds overlaying the Great Dark Spot at its southern and northeastern boundaries may have been analogous to lenticular clouds that form over mountains on Earth. While reading this description, it is fair to question my use of the past tense. This is because the Great Dark Spot has vanished.

Images taken in 1994 by the Hubble Space Telescope confirm the disappearance of the Great Dark Spot. The second, smaller dark spot, DS2, that was seen during the Voyager 2 encounter was also missing. The absence of these dark spots was surprising. The Hubble Wide-Field Planetary Camera science team initially assumed that the two storm systems might be near the edge of the planet's disk, where they would not be particularly obvious. This was not the case.

An analysis of image coverage revealed that less than 20° of Neptune's longitude had been missed. The Great Dark Spot covered almost 40° of longitude at the time of the Voyager 2 encounter. Even if it were on the edge of the disk, it would appear as a "bite" out of the limb, however, no such feature was detected. Recent ground-based observations confirm this finding.

Less than one year later, the Hubble Space Telescope discovered a new Great Dark Spot, located in the northern hemisphere of the planet. Because the planet's northern hemisphere is now tilted away from Earth, the new feature appears near the limb of the planet. The spot is a near mirror-image of a similar southern hemisphere dark spot that was discovered by Voyager 2.

Like its predecessor, the new spot has high-altitude clouds along its edge, caused by gases that have been pushed to higher altitudes where they cool to form methane ice crystal clouds. The Dark Spot may be a zone of clear gas that is a window to a cloud deck lower in the atmosphere. How "Great" this Dark Spot turns out to be will depend upon its longevity. Schedulers indicate that the Hubble Space Telescope will continue to monitor Neptune's atmosphere from time to time.

Most of the winds on Neptune blow in a westward direction, which is retrograde, or opposite to the rotation of the planet. Near the position of the 1995 Great Dark Spot, Voyager 2 measured retrograde winds blowing up to 2400 km/hr – the strongest winds measured on any planet. The temperature difference between Neptune's strong internal heat source and its frigid cloud tops might trigger instabilities in the atmosphere that drive these large-scale weather changes. In addition, since Neptune is so cold, most astronomers believe that there is much less frictional resistance within the atmosphere. Simply put, once a wind begins to blow, there is not much in the way to slow it down.

Triton

Triton is Neptune's largest moon. It was discovered by the British astronomer William Lassell scarcely a month after the discovery of Neptune, in 1846.

Triton is an oddity among moons in that its orbit is highly tilted to the plane of Neptune's equator (156.8°), and it is in a retrograde orbit. It is the only large satellite in the solar system to circle a planet in a retrograde direction. These facts have led scientists to believe that Triton formed independently of Neptune and was later captured by Neptune's gravity. If that is the case, tidal heating could have melted Triton in its originally eccentric orbit, and the satellite might even have been liquid for as long as one billion years after its capture by Neptune. It is the second most distant of Neptune's satellites, lying some 355 000 km from the planet. The diameter of Triton is 2705 km.

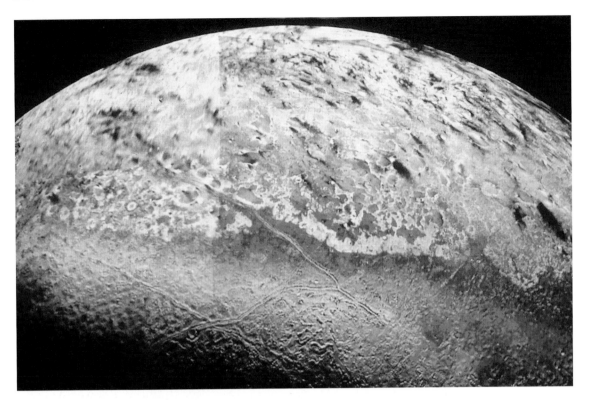

Triton. (NASA)

The pink hue of Triton is thought to result from a slowly evaporating layer of nitrogen ice. Some of the landscape features a cantaloupe-textured terrain encompassing a region roughly 1000 km across. Complex tectonic and volcanic forces involving icy viscous fluids combined to develop the deformed pattern of this landscape.

Triton's surface is covered with a thin layer of nitrogen and methane ice. Most of the geologic structures on Triton's surface are likely to be formed of water ice, because nitrogen and methane ice are too soft to support their own weight. Dark streaks across the south polar cap are the result of recent geyser-like eruptions of gas, dust and ice venting from beneath the cap into the satellite's near-vacuum atmosphere. Nitrogen and methane, which form a thin veneer on Triton, turn from ice to gas at less than 100 K. Most of the geologically recent eruptions at the low temperatures found on this moon are due to the nitrogen and methane. The fine particles within the eruptions are carried to altitudes of 2–8 km and then blown downwind before being deposited on Triton's surface.

An extremely thin atmosphere extends as much as 800 km above Triton's surface. Tiny nitrogen ice particles may form thin clouds a few kilometers above the surface. Triton is very bright, with an albedo of 0.60–0.95, depending on the surface composition.

Eight planets; the first depiction of Neptune as anything but a dot on a map. Flammarion, Camille, "Les conditions de la vie dans l'univers" *L'Astronomie*, **4**:161–74, Paris, 1886. (Linda Hall Library)

The atmospheric pressure at the surface of Triton is about 1.4 pascals. That is equal to 1/70000 the sea level atmospheric pressure on Earth. The temperature at Triton's surface is a mere 38 K, the coldest surface of any body yet visited in the solar system. At 800 km above the surface, the temperature is 95 K.

Interesting facts

Neptune is slightly more than 17 times as massive as the Earth.

Voyager 2 measured the surface temperature of Triton at 38 K ± 4. It is the coldest body so far observed in the solar system.

Neptune's orbit is the second least eccentric among planets, varying from circular by less than 1%.

Neptune, showing rings! Breen, James, *The Planetary Worlds*, London, 1854. (Linda Hall Library)

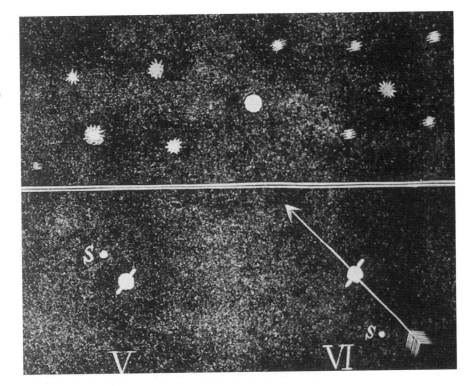

When it was discovered, Neptune was within the boundaries of the constellation Aquarius. Since its discovery, Neptune has not made one complete orbit around the Sun. On 8 June 2011 Neptune will finally complete one post-discovery orbit.

Nereid's orbit is the most eccentric in the solar system, with an eccentricity of 0.7512. Its distance from Neptune varies from 1 353 600 km to 9 623 700 km.

The Sun is 1096 times as bright from the Earth as from Neptune.

The accuracy of Voyager 2's arrival at Neptune has been compared to sinking a golf shot from a distance of 3637 km, though Voyager 2 did take advantage of a few minor course corrections along the way.

Neptune's volume is 59 times as great as the Earth's.

The natural satellite with the shortest orbital period is Naiad. It orbits Neptune in only 7 hours and 4 minutes.

The fastest winds in the solar system have been measured on Neptune as 2400 km/hr.

Neptune, with an average apparent motion (against the background of stars) of only 22 seconds of arc per day, takes approximately 85 days to traverse a distance equal to the width of the Full Moon.

If Neptune had no internal heat source, its temperature would be approximately $-220\,°C$, rather than the measured $-216\,°C$.

Observing data

Future dates of conjunction

24 Jan 2000
26 Jan 2001
28 Jan 2002
30 Jan 2003
2 Feb 2004
3 Feb 2005
6 Feb 2006
8 Feb 2007
11 Feb 2008
12 Feb 2009
15 Feb 2010

Future dates of opposition

26 Jul 1999
27 Jul 2000
30 Jul 2001
2 Aug 2002
4 Aug 2003
6 Aug 2004
8 Aug 2005
11 Aug 2006
13 Aug 2007
15 Aug 2008
18 Aug 2009
20 Aug 2010

Recent data

All recent data involving Neptune is a direct result of the flyby, in August 1989, of the Voyager 2 spacecraft. Passing about 4950 km above Neptune's north pole, Voyager 2 made its closest approach to any planet since leaving Earth in 1977. Five hours later, Voyager 2 passed about 40 000 km from Neptune's largest moon, Triton, the last solid body the spacecraft had an opportunity to study.

The most obvious feature of the planet in the Voyager pictures is its blue color, the result of methane in the atmosphere. Methane absorbs the longer, reddish wavelengths of sunlight, leaving the shorter, bluish wavelengths to be reflected back into space.

While methane is not the most abundant gas in Neptune's atmosphere, it is probably the most important. Scientists now have a better understanding of the way methane cycles through the atmosphere of Neptune.

Initially, solar ultraviolet radiation destroys methane high in Neptune's atmosphere by converting it to hydrocarbons and haze particles of more complex polymers. The haze particles sink to the cold lower stratosphere, where they freeze. In a phenomenon which must resemble snow on the Earth, the hydrocarbon ice particles gently fall into the warmer troposphere, where they evaporate back into gases. These gases migrate deeper into the atmosphere where the temperature and pressure are higher. There, they mix with hydrogen and regenerate methane. Buoyant, convective clouds of methane then rise to the stratosphere, returning methane to that part of the atmosphere. It is important to note that there is no net loss of methane in this process.

The atmosphere of Neptune is marked by a wide variety of cloud types. For more information about the clouds of Neptune see the section entitled The Great Dark Spot.

Voyager 2 also measured the amount of heat radiated by Neptune's atmosphere. The atmosphere above the clouds is hotter near the equator, cooler in the mid-latitudes and warm again at the south pole. Temperatures in the stratosphere were measured to be 750 K, while at an altitude where the atmospheric pressure equaled 10 133 pascals, they were measured to be 55 K. The primary heat source is convection in the atmosphere. Such motion results in compressional heating. The cycle begins in the mid-latitudes where gases rise and cool, they then drift toward the equator and the pole, where they sink and are warmed.

Most of the winds on Neptune blow in a westward direction, which is opposite to the rotation of the planet. Near the Great Dark Spot, there are winds blowing up to 2400 km/hr. These are the fastest winds in the solar system.

Neptune's magnetic field is tilted 55° from the planet's rotational axis, and is offset at least one-quarter the diameter of Neptune itself (12 400 km) from the physical center. The field strength at the surface varies, depending on which hemisphere is being measured, from a maximum of more than 0.0001 tesla in the southern hemisphere to a minimum of less than 0.00001 tesla in the northern. Because of its unusual orientation and the tilt of the planet's rotation axis, Neptune's magnetic field goes through dramatic changes as the planet rotates in the solar wind.

There are many similarities between Neptune's magnetic field and that of Uranus, which is tilted 59° from the rotation axis, with a center that is offset by 0.3 Uranus radii. Scientists have no definite answers yet, but think that the tilt may be characteristic of flows in the interiors of both Uranus and Neptune and unrelated to either the high tilt of Uranus's rotation axis or possible field reversals at either planet. Neptune's magnetic field polarity is the same as those of Jupiter and Saturn, and opposite to that of Earth.

Details of Neptune's magnetic field provided another clue to the planet's structure and behavior. Observers on Earth had not been able to determine the exact length of a Neptunian day. Cloud motions are a poor indicator of the rotation of the bulk of the planet, since they are affected by strong winds and vary substantially with latitude. The best telescopic estimate was a rotation period of approximately 18 hours. The best indicator of the rotation period of the planet beneath the clouds is periodic radio waves generated by the magnetic field. Voyager's planetary radio astronomy instrument measured these periodic radio waves, and determined that the rotation rate of the main bulk of Neptune is 16 hours 6.6 minutes.

Voyager 2 detected auroras, similar to those on Earth, in Neptune's atmosphere. The auroras on Earth occur when energetic particles of the solar wind strike the atmosphere as they spiral down the magnetic field lines. But because of Neptune's complex magnetic field, its auroras are extremely complicated processes that occur over wide regions of the planet, not just near the planet's magnetic poles. Because of Neptune's vast distance from the Sun, the auroral power on Neptune is weak, estimated at about 50 million watts, compared to 100 billion watts on Earth.

In the 1980s, there were several occultations of stars by Neptune. Such events are important because observers are able to analyze the starlight and how it changes as it passes through the upper layer of Neptune's atmosphere. Thus, clues to the structure of the atmosphere may be obtained.

During nearly every occultation, the star's light was extinguished and then re-appeared before Neptune passed in front of it. Astronomers concluded that some material orbits Neptune, and is responsible for the occasional blockage of the star's light. A ring was postulated. In each observed event, however, astronomers saw that the ring or rings did not appear to completely encircle the planet – rather, each appeared to be an arc segment of a ring.

According to theory, with no external force acting upon them, rings must orbit a planet at about the same distance from the center all the way around. Ring material, if unrestrained, will tend to disperse uniformly around the planet. In order to have arcs instead of rings, scientists thought that additional bodies, perhaps small shepherding satellites, must keep the arcs in their place by gravity.

When Voyager 2 was close enough, its cameras photographed three bright patches that looked like ring arcs. But closer approach, higher resolution and more computer enhancement of the images showed that the rings do, in fact, go all the way around the planet. The rings are so diffuse, and the material in them so fine, that Earthbound astronomers simply hadn't been able to detect the full rings.

Rings of Neptune

Ring	Distance (km)[a]	Width (km)	Designation
Diffuse	41 900	15	1989N3R
Inner	53 200	15	1989N2R
Plateau	53 200	5800	1989N4R
Main	62 930	50	1989N1R

Note:

[a] Distance is measured from Neptune's center to the ring's inner edge.

The rings of
Neptune as imaged
by Voyager 2.
(NASA)

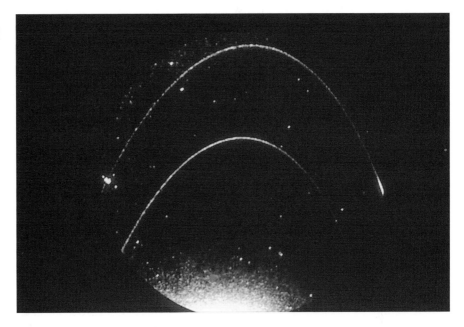

The Main ring contains three separate regions where the material is brighter
and denser, and this explains most of the sightings of "arcs." Some scientists
suspect that the Diffuse ring may extend all the way down to Neptune's
cloud tops. The material varies considerably in size from ring to ring. The
largest proportion of fine material – approximately the size of smoke particles – is in the Plateau ring. All other rings contain a greater proportion of
larger material.

Satellites

	Size (km)	Mass (kg)[a]	Density (g/cm^3)	Orbital Period	Eccentricity	Inclination	Distance from planet (km)
Naiad	58	—	—	07^{h}03.9^m	0.00	4.74°	4.8227 10^4
Thalassa	80	—	—	07^{h}28.5^m	0.00	0.21°	5.0075 × 10^4
Despina	148	—	—	08^{h}01.9^m	0.00	0.07°	5.2526 × 10^4
Galatea	158	—	—	10^{h}17.4^m	0.00	0.05°	6.1953 × 10^4
Larissa	208(×178)	—	—	13^{h}18.7^m	0.00	0.20°	7.3548 × 10^4
Proteus	436(×416×402)	—	—	01^{d}02^{h}56.1^m	0.00	0.55°	1.17647 × 10^5
Triton	2705	2.147 × 10^{22}	2.054	05^{d}21^{h}02.7^m (r)[b]	0.00	156.834°	3.5476 × 10^5
Nereid	340°	—	—	360^{d}03^{h}16.11^m	0.7512	7.23°	5.5134 × 10^{6c}

Notes:
[a] Satellites with no value for their mass have never had this quantity measured to any degree of accuracy.
[b] Retrograde.
[c] Average.

Historical timeline

28 January 1613	Italian astronomer Galileo Galilei made an observation of Neptune when it was near Jupiter, but mistook it for a fixed star; he may also have seen it on 28 December 1612.
19 September 1702	At 13^h26^m ET Jupiter occulted Neptune.
23 September 1846	German astronomer Johann Gottfried Galle, working at the Berlin Observatory, discovered Neptune within the boundaries of the constellation Aquarius.
10 October 1846	British astronomer William Lassell discovered Triton.
1949	Dutch-born US astronomer Gerard Peter Kuiper discovered Nereid.
1989	Scientists with US space probe Voyager 2 discovered Despina, Galatea, Naiad and Thalassa. Stephen Synnott discovered Larissa and Proteus.
24 August 1989	US space probe Voyager 2 flew within 4950 km of the cloud tops of Uranus (Voyager 2 was launched 20 August 1977).

Pluto

Physical data

Size   2320 km

Mass 1.290×10^{22} kg

Escape velocity   1.22 km/s (4392 km/hr)

Temperature range

	Minimum	Average	Maximum
	$-240\,°C$	$-229\,°C$	$-218\,°C$

Oblateness   0

Surface gravity $0.4\ m/s^2$

Volume $6.545 \times 10^9\ km^3$
0.602% that of Earth

Magnetic field strength and orientation   Pluto has no known magnetic field. It is almost certainly too small to maintain a molten metallic core, a prerequisite for the development of a magnetic field.

Albedo (visual geometric albedo)   0.30

Density (water = 1) $2.05\ g/cm^3$

Solar irradiance $0.9\ watts/m^2$

Atmospheric pressure = 0   Pluto may develop an incredibly thin atmosphere when near perihelion, but normally its atmosphere lies frozen on the surface.

Composition of atmosphere   Methane is the only component which has been detected. Other gases, such as nitrogen, may exist in frozen form.

Wind speeds   For much of its orbit, Pluto is too far from the Sun (and thus too cold) to maintain a gaseous atmosphere. Near perihelion, some atmosphere may develop due to solar heating, but the measurement of wind speeds on Pluto (if any) remains beyond current technology.

Outstanding surface features

The Hubble Space Telescope photographed Pluto in the late-1990s. Images show that Pluto has a very heterogeneous surface. Examination of images revealed that Pluto's surface has more large-scale contrast than any planet except Earth. Hubble found 12 major regions on the surface differentiated by a high or low reflectivity. The resolving power of the Space Telescope at the distance of Pluto is approximately 161 km per pixel.

Thanks to the Hubble Space Telescope, planetary scientists were able to construct the first map of the surface of Pluto. It covers 85% of the planet's surface and confirms that Pluto has a dark equatorial belt and bright polar caps.

Orbital data

Period of rotation   6.3872 days
153.2928 hours
$6^d9^h17.6^m$

Period of revolution (sidereal orbital period)   248.54 years
90777.3 days
$248^y197^d5.5^h$

Synodic period $366^d17^h31.2^m$

Equatorial velocity of rotation   47.6 km/hr

Velocity of revolution   4.74 km/s (17064 km/hr)

Distance from Sun   Average   39.5294 AU
5913520000 km
Maximum   50.299 AU
7524587000 km
Minimum   29.647 AU
4435128000 km

Distance from Earth   Maximum   51.30 AU
7676691980 km
Minimum   28.63 AU
4283023020 km

Apparent size of Sun (average)   0.013°

Apparent brightness of Sun $m_{vis} = -18.5$

Orbital eccentricity   0.2482

Orbital inclination   17.148°

Inclination of equator to orbit   122.52°

Observational data

Maximum angular distance from Sun   180°

Brilliancy at opposition   Maximum   +13.6
Minimum   +15.95

Angular size*   Maximum   0.11″
Minimum   0.065″

* This measurement is the apparent angular diameter of Pluto, measured in seconds of arc, as seen from Earth.

Discovery

The discovery of Pluto was made by Clyde William Tombaugh. But Tombaugh did not merely travel to a dark site, set up a telescope, plop in an eyepiece and exclaim, "There it is!" No, Tombaugh's detection of the solar system's ninth major planet was a result of detailed procedures, exceptional tenacity and specific direction. This direction, incidentally, started with none other than Percival Lowell.

As early as 1902, Lowell was lecturing and writing about his conviction that a planet existed beyond the orbit of Neptune. He had studied the orbits of a number of comets and had noticed gaps and groupings which he felt were significant. In 1905, the first of his two extended searches for "Planet X" began. This phase lasted until 1909, and was essentially carried out by Lowell, a trio of graduate assistants and William T. Carrigan, a "computer" from the US Naval Observatory's Nautical Almanac Office in Washington, DC. Lowell employed Carrigan to perform a number of calculations on the orbits of Uranus and Neptune. Lowell was interested in "residuals," differences between the calculated positions of the planets and the positions which were actually observed.

Lowell's second search spanned the years 1910–1915. This search was based upon calculations using the formulae of celestial mechanics. In essence, it was similar to the predictions of Adams and LeVerrier, but using residuals which were much smaller. The calculations, thus, were much more difficult and involved. This search culminated in the publication entitled "Memoir on a Trans-Neptunian Planet," in *Memoirs of the Lowell Observatory*, vol. 1, 1915.

At about the same time as Lowell's observations were being carried out, William Henry Pickering of the Harvard College Observatory made a series of published predictions of trans-Neptunian planets based upon his own calculations. The first report appeared in 1908. He called the object "Planet O," and gave it a distance from the Sun of 51.9 astronomical units, a period of 373.5 years, a mass twice that of Earth and an estimated visual magnitude of 11.5–13.4. In 1911 Pickering published three additional predictions, dubbed, invariably, Planets P, Q and R. These objects were postulated to be quite distant. P was 123 astronomical units from the Sun, Q was 875 astronomical units distant, and R was at the incredible distance of 6250 astronomical units. In addition, Planet Q was given a mass by Pickering of 20 000 times that of Earth. Such an object would have made quite a companion to the Sun.

In 1928, Pickering published a revised set of orbital elements for Planet P and also a brand new prediction of Planet S. This object was at a more realistic distance of 48.3 astronomical units, with an orbital period of 336 years, a mass 5 times that of Earth and an estimated visual magnitude of 15. None of Pickering's planets were ever found, but the nearly continuous string of predictions does indicate that a trans-Neptunian planet was not far from the minds of astronomers.

At Lowell Observatory, the third search for Planet X began in 1927. Plans were formulated and money was secured for a new photographic telescope to better continue the hunt. The 33-cm A. Lawrence Lowell Astrographic

Telescope took its first exposure in the hunt for the trans-Neptunian planet on 6 April 1929. It was an hour-long exposure centered on the star δ Cancri. The man at the telescope was Clyde Tombaugh.

Tombaugh had been corresponding with Vesto Melvin Slipher, Director of the Lowell Observatory, during 1928. Early in 1929, Slipher offered Tombaugh a position to work with the 33-cm Astrograph. Tombaugh traveled from Burdette, KS, to Flagstaff, AZ, arriving at Lowell Observatory on 15 January 1929, with absolutely no idea that he would be conducting a search for a new planet.

Tombaugh exposed more than 150 plates before his epic exposures on 23 January 1930 and 29 January 1930. Even then, the discovery was not immediate. To examine the plates, Tombaugh used the observatory's blink comparator, where first one plate is illuminated and then the other. A small portion of the plates is studied with a microscope incorporated into the apparatus. The necessity of blinking other plates meant Tombaugh did not examine the exciting pair until several weeks after they were exposed. Then, according to Tombaugh's notes, on 18 February 1930 at 4 p.m., Planet X was discovered on the comparator.

V. M. Slipher was a cautious man. He wanted confirmation that the object identified by Tombaugh was indeed Percival Lowell's long-anticipated Planet X. More exposures were taken through all the telescopes at the observatory. A determination of the orbit was deemed necessary, however the staff at the observatory at that time did not have the expertise to calculate a precise orbit so Slipher enlisted the aid of two non-Lowell astronomers. Time moved on, but still no announcement was made. In retrospect, this was partly due to Slipher's desire to be "on the mark" regarding the discovery of a new planet. Slipher, better than most, was well acquainted with the reputation that Lowell Observatory had received due to Percival Lowell's obsession with life on Mars. But Slipher also wanted the opportunity for Lowell Observatory to observe the newest addition to the solar system for as long as possible without competition.

Finally, on 13 March 1930, on what would have been the seventy-fifth birthday of Percival Lowell, the announcement was made. It is interesting to note that this was also the date of the first planetary discovery since antiquity, by Sir William Herschel, 149 years earlier. The Lowell Observatory's reputation was secure and Clyde Tombaugh – who less than two years before his discovery was a farmboy on the plains of Kansas – was assured a place in history.*

---

* On a personal note, this writer would like to lament the passing on 17 January 1997 of Clyde William Tombaugh, whom I knew. Clyde was one of only four people in the vast history of our solar system to have discovered or predicted a major planet. Beyond the status of a legend, though, Clyde Tombaugh was most enjoyable. He was always willing to discuss observing, telescopes or any other aspect of astronomy, including, "his planet." Clyde told a great story and had a wonderful sense of humor. His grasp of facts and figures was amazing and remained so his entire life. Each visit with Clyde (and his wife Patsy) was a learning experience and even though we saw each other infrequently, we could always pick right up where we left off. I last saw Clyde four weeks before his death and he was as sharp and as funny as ever. He will be missed by many.

Important concepts

Is Pluto a planet, an asteroid or a comet?

On March 13 1930, astronomers announced Clyde Tombaugh's discovery of a planet that the International Astronomical Union initially called "Object Lowell Observatory." Around 1980, a few people started to call for a re-evaluation of Pluto's status. Perhaps it is too small to be a major planet, they suggested; instead, it should be considered a large asteroid.

There are many lines of reasoning in the determination of Pluto's status in the solar system. Some astronomers have stated that the only reason that Pluto is considered a major planet was because of Percival Lowell and the search for Planet X. When Pluto was finally discovered, the media barrage that ensued, and the response to it by the Lowell Observatory, virtually guaranteed the classification of Pluto as a major planet.

Those who oppose labeling Pluto as a planet often cite its small size. Pluto is smaller than seven moons of the solar system. It must be stated, however, that Mercury (a major planet) is smaller than both Ganymede and Titan.

The solar system seems to be divided into two realms, dominated by the terrestrial planets near the Sun and the Jovian planets further out. These groups have similar characteristics and govern the regions of the solar system in which they are found. Pluto, it is pointed out, does not seem to fit into either group.

A decision which has persuaded many astronomers that Pluto is not a major planet was made by the Minor Planet Center of the International Astronomical Union. For orbital computations involving solar system objects, this group now uses the largest asteroid, Ceres, instead of Pluto as a ninth perturbing body.

Pluto's mass is so tiny that it has no appreciable effect on the motion of Uranus. A study confirming this was published by Ernest William Brown within a year of Pluto's discovery. Whether the gravitational attraction of Pluto affects Neptune is not known. More observations will need to be made to determine if there are small perturbations.

Pluto's orbit is different from those of the eight other planets. It has a high inclination (17°) and a large eccentricity (0.2482) that some astronomers contend are actually more typical of asteroid or periodic-comet orbits. (Again, it must be stated that Mercury's eccentricity (0.2056) is nearly as great as that of Pluto.) These astronomers argue that Pluto seems more akin to the many small bodies which exist in the outer realms of the solar system.

Our knowledge of these small bodies began to increase in 1977, with Charles Kowal's discovery of a Saturn-crossing object, now known as Chiron,* whose orbit ranges from 8.5 astronomical units at perihelion to about 19 astronomical units at aphelion.

Complicating (or perhaps, refining) matters is the recent discovery of a number of trans-Neptunian objects that are asteroidal in appearance. It is still uncertain whether any of these are similar to Pluto in that they cross the orbit of Neptune. Astronomers continue to investigate the objects already discovered and to add to the inventory of this part of the solar system.

* Chiron, the asteroid, should not be confused with Charon, the moon of Pluto.

Some planetary scientists think that Pluto is unlikely to be a unique object that ended up by chance in the stable orbit it now occupies. Pluto's orbit exhibits a degree of gravitational locking, as it is in a 2:3 resonance with the orbit of Neptune. The existence of a satellite around Pluto may strengthen this hypothesis, since if there were only two such objects (Pluto and Charon), the odds of their meeting would have been small.

Ceres, an asteroid with a diameter of 940 km, has been shown recently to have water ice on its surface. Chiron has exhibited something resembling a cometary coma and tail. Several anti-planet proponents have pointed out that Pluto, along with the other very distant objects would likely have comas (and even tails) if brought within a few astronomical units of the Sun. This argument breaks down, however, as most of the icy moons of the giant planets – and even the Jovian planets themselves – would exhibit the same phenomena if their solar distances were somehow decreased.

The existence of the moon Charon has been cited as sufficient grounds for calling Pluto a major planet. Detractors point out that the Pluto–Charon mass ratio is only 7:1. Compare this to the 82:1 ratio for the Earth–Moon system. A comet which had split into pieces could be quite long lived, especially at Pluto's distance. And there are other non-planet binary objects. The asteroid Ida has been imaged with a second, smaller asteroid seemingly in orbit around it.

US astronomer S. Alan Stern, in 1993, proposed three classification standards regarding bodies labeled "major planet:"

(1)   A major planet must orbit a star directly.
(2)   Upper and lower size limits can be based entirely on physical reasoning. For the upper limit, a planet cannot sustain fusion of any sort, and for the lower limit, a major planet should approximate hydrostatic equilibrium – it should be round.
(3)   Lastly, a major planet must be a naturally occurring object.

The suggestions of Stern are based solely on objective physical characteristics. Notice that these criteria do not account for detectable gravitational perturbations a planet might have on other bodies. Thus, an Earth-sized planet, situated several hundred astronomical units from the Sun, probably would not have discernible effects on objects in the inner solar system, but would not be excluded from receiving the designation of major planet.

Canadian-born comet hunter David H. Levy wonders what we would call Pluto should it lose its planetary status. An asteroid? Levy points out that Pluto has a detectable atmosphere, at least near perihelion, a trait not shared with any asteroid. A comet? Levy knows comets. He contends that Pluto is much larger in diameter than any other cometary object.

Planetary scientist Larry A. Lebofsky has suggested that we could simply define Pluto's diameter as the smallest acceptable size for a major planet. Smaller bodies, like those recently discovered near and beyond Neptune's orbit, would get asteroid designations. Or, as another astronomer has suggested, we could adopt a specific diameter, say, 1000 km, as a dividing line. Ceres, an asteroid, is smaller than this value, and Pluto, a major planet, is larger.

It may be that the most compelling reason for Pluto's classification as a major planet is the way in which it has been viewed since its discovery. As an educator, I well remember the concerned questions of teachers and students when this debate surfaced in the media during the mid-1990s. We all love Pluto. This may be due to its diminutive size, its extremes of distance and temperature, or even to the popularity of Walt Disney, Inc. Professional astronomers who desire a de-classification of Pluto as a major planet would be well advised to consider the impact such a change would have on the public at large.

The story of Pluto's atmosphere

Pluto's orbit was determined soon after its discovery in 1930. It was shown to have both a surprisingly high eccentricity (0.2482) and the highest inclination of any planet (17.148°). Pluto's rotation rate (6.3872 days) was first accurately measured in 1954, but the breakthrough that first gave an idea of conditions on Pluto was the discovery of methane (CH_4) in 1976.

Spectroscopic studies of Pluto showed a solid methane absorption feature in its spectrum. This indicated that Pluto was colder than about 55 K, otherwise the methane would not be in the solid phase. It also implied that Pluto was a relatively bright, frost-covered planet. This was important in the determination of the size of Pluto. We now know that Pluto is a small planet with a bright surface as opposed to a large planet with a darker surface.

One fascinating aspect of Pluto's atmosphere occurs when the planet nears perihelion. At that time, the surface temperature rises just enough to evaporate some of the ices and to generate an atmosphere. Observations of Pluto during stellar occultations suggest that the surface pressure during times of maximum atmosphere is 10 pascals. Higher-dispersion spectroscopy has also revealed the possible presence of nitrogen (N_2) in the atmosphere.

Spectroscopic studies of Charon show the presence of water ice, rather than methane. It is possible that Charon's "atmosphere," like Pluto's once contained methane, but being of much smaller mass than its parent planet, Charon's gravitational attraction was not enough to keep the methane from escaping into space during the warm times near perihelion.

Charon; its discovery from Earth

The date was 22 June 1978. At the US Naval Observatory in Washington, DC, US astronomer James Christy was working on improving the orbital parameters of Pluto. After the careful study of a number of photographic plates, Christy noticed that the image of Pluto seemed lumpy. There was an irregular blob that seemed to be attached to Pluto. Christy wondered about the focus, but the other stars on the plates were round.

In addition, the blob seemed to move around Pluto with a period of approximately 6.4 days. There were even times when the blob passed in front of Pluto. It was this circumstance which allowed the orbital period to be so quickly established. Each time Charon passed in front of (or behind) Pluto, the combined light of the system would dim slightly. Soon after Christy's discovery, the object was confirmed as a satellite. It was given the name Charon, the ferryman who, in Roman mythology, transfers the souls of the dead across the river Styx to the underworld realm of the god Pluto.

The tilt of Pluto's axis helped make this discovery possible. Most planets have north poles aligned roughly perpendicular to their orbit planes. The exceptions are Uranus, whose axis is tilted 97.86° and Pluto, whose axis is tilted even more, 122.52°. As fortunate happenstance (luck) would have it, during the period from 1985 to 1990, Pluto's equator and Charon's orbital plane were aligned with the line-of-sight of observers on the Earth. It was, therefore, easy to compute the revolution of Charon. Since the discovery, Charon's orbital period has been refined to 6.387 days.

Are there planets in our solar system beyond Pluto?

The search for a tenth planet has been going on since Clyde Tombaugh first spotted the images of Pluto in 1930. The list of observers who have laid claim to spotting or photographing "Planet X" (now that Pluto no longer fits that description) would fill this volume. The following examples are but a few of the more respected attempts.

Scarcely more than a month after Pluto was announced, R. M. Stewart, the Director of the Dominion Observatory in Ottawa, Canada, announced that two images of a possible tenth planet had been located while observers there were searching for pre-discovery images of Pluto. (Such images would be helpful in refining the planet's orbit.) The British Astronomical Association even went so far as to publish an orbit, giving the object a distance of approximately 5.96 billion km, or around 40 astronomical units. In the many years since the announcement, no further images of this supposed planet have ever been seen.

In 1950, K. Schutte of Munich used data from eight periodic comets to suggest the orbit of a tenth planet at 11.5 billion km, or 77 astronomical units. Four years later, H. H. Kitzinger of Karlsruhe, using the same eight comets, extended and refined the work, placing the supposed planet at a distance of 65 astronomical units with a visual magnitude of 11. In 1957 Kitzinger reworked the problem and, after unsuccessful photographic searches, he reworked the problem once again in 1959 – all for naught. No such planet has ever been found.

Halley's Comet has also been used as a "probe" for trans-Plutonian planets. In 1942 R. S. Richardson found that an Earth-sized planet at 36.2 astronomical units would delay Halley's perihelion passage so that it agreed better with observations. In 1972, Brady predicted a planet at 59.9 AU with a size about as big as Saturn. It was believed that such a planet would reduce the differences between the predicted positions of Halley's Comet and the actual observations. This gigantic planet was also searched for, but never found.

Tom van Flandern examined the positions of Uranus and Neptune in the 1970s. He found that the observations of Neptune could be made to fit a calculation of its orbit for only a few years at a time and then started to drift away. He could make the calculations for Uranus fit for any one orbit, but not the following one. In 1976, van Flandern became convinced that there was a tenth planet. After the discovery of Charon in 1978 showed the mass of Pluto to be much smaller than expected, van Flandern began to investigate the Neptunian satellite system. van Flandern thought the tenth planet had formed beyond Neptune's orbit. He suggested that the tenth planet might be near aphelion in a highly elliptical orbit. If the planet is dark, he suggested, it might be as faint as visual magnitude 16 or 17.

In 1987, John Anderson at the Jet Propulsion Laboratory in Pasadena, CA, examined the motions of the two spacecraft Pioneer 10 and Pioneer 11, to see if any deflection due to unknown gravity forces could be found. None was found, yet Anderson concluded that a tenth planet may still exist. Anderson deduced that the tenth planet must have a highly elliptical orbit, carrying it far away to be undetectable now but periodically bringing it close enough to leave its disturbing signature on the paths of the outer planets. He suggests a mass of five Earth masses, an orbital period of 700 to 1000 years and a highly inclined orbit. Its perturbations on the outer planets should not be detected again until the year 2600. Anderson hoped that gravitational effects slightly altering the paths of the two Voyager spacecraft would help to pin down the location of this planet. To date no evidence has been found.

In 1977–1984 Charles Kowal performed a new systematic search for undiscovered bodies in the solar system, using Palomar Observatory's 122-cm Schmidt telescope. During his search, Kowal found 5 comets and 15 asteroids, including Chiron, the most distant asteroid known when it was discovered. Kowal also recovered 4 lost comets and one lost asteroid. Kowal did not find a tenth planet, and concluded that there was no unknown planet brighter than 20th magnitude within 3° of the ecliptic.

Chiron was first announced as a "tenth planet," but was immediately designated as an asteroid. Kowal suspected it to be very comet-like, and later it indeed developed a short cometary tail. In 1995, Chiron was classified as a comet.

In 1992, an even more distant asteroid was found, Pholus. Later in 1992, an asteroid outside Pluto's orbit was found, followed by five additional trans-Plutonian asteroids in 1993, and at least a dozen in 1994. More are discovered each year.

Meanwhile, the spacecraft Pioneers 10 and 11 and Voyagers 1 and 2 had traveled outside the solar system, and could also be used as "probes" for unknown gravitational forces possibly from unknown planets – nothing has been found.

However, the Voyagers did provide a set of definitive facts by accurately measuring the masses of the outer planets. When these updated masses were inserted in the numerical integrations of the solar system, the perceived errors in the positions of the outer planets finally disappeared. It seems that the search for Planet X finally has come to an end. There was no "Planet X." Instead an asteroid belt outside the orbit of Pluto was found.

Clyde Tombaugh, discoverer of Pluto, once stated that he was convinced that no planet brighter than visual magnitude 16 exists, and that he would even go so far as to claim that none brighter than magnitude 17 would be found. Tombaugh himself spent 13 additional years examining over 90 million images of approximately 30 million stars – and this was *after* the discovery of Pluto.

For the still skeptical, in January 1983, the Infrared Astronomical Satellite (IRAS) – a collaborative effort by the United States (NASA), the Netherlands (NIVR) and the United Kingdom (SERC) – was launched to image the sky at infrared wavelengths. The survey covered the sky in overlapping strips as the satellite scanned each orbit. Most (96%) of the sky was covered by at least two separate scans and 2/3 of the sky was also covered by a third.

IRAS contained a liquid helium-cooled 0.6-m Ritchey–Chretien telescope. It conducted an all-sky survey at wavelengths ranging from 8 to 120 microns in four broadband photometric channels. Some 250 000 point sources were detected. IRAS also made pointed observations of selected objects with integration times lasting up to 12 minutes, providing up to a factor of 10 increase in sensitivity relative to that of the rest of the survey.

Within the data reduction programming was a subroutine to specifically identify moving objects. It seemed that some of the project scientists had hopes of locating Planet X. Several comets (including the near-Earth approaching IRAS–Araki–Alcock in 1983) and a number of asteroids were discovered this way. But no planets. Although possible, it seems unlikely that IRAS could have missed a planet-sized object.

Interesting facts

On average, the Sun appears 1905 times fainter at Pluto than it does from Earth.

Although the discovery of Pluto was not announced until 13 March 1930, the two plates that were compared to detect its motion were taken on 23 January and 29 January 1930.

At the time of its discovery, Pluto was within the boundaries of the constellation Gemini.

On 1 January 2000, Pluto will have traveled 28.14% of its orbit around the Sun since its discovery.

Pluto will not complete one post-discovery orbit until 8 August 2178.

Even though it takes Pluto more than 248 years to orbit the Sun, it still travels, on average, at a respectable 17 064 km/hr.

Pluto and Neptune will never collide. They can never be less than about 386 000 000 km from each other.

Pluto may appear within the boundaries of 41 constellations.

The escape velocity of Pluto is only 11% that of Earth.

The Earth is 463 times as massive as Pluto.

On average, it takes the light of the Sun 5½ hours to reach Pluto. Because of Pluto's varying distance from the Sun, this time can be as short as 4 hours 6 minutes, when Pluto is at perihelion, or as long as 6 hours 58 minutes, when it is at aphelion.

The surface gravity of Pluto is only 4.1% that of the Earth.

The synodic period of Pluto is only 35^h42^m more than one Earth year.

We often think of Uranus when we imagine a planet "tipped over on its side." However, Pluto's axial tilt is more than 24° greater than that of Uranus.

The volume of the Earth is 159 times that of Pluto.

Pluto was discovered with the A. Lawrence Lowell Astrographic refractor. This telescope had a 33-cm, Cooke type, three-element lens with a focal length of 1.69 m. This meant that on a photographic plate 35.6 cm × 43.2 cm a field of view 12° × 14° could be captured, with a scale of 122 seconds of arc per millimeter.

As seen from the Earth, Pluto moves very slowly. Its average apparent motion (against the background of stars) is a leisurely 14 seconds of arc per day. This means that it takes Pluto roughly 130 days to travel a distance equal to the width of the Full Moon.

Observing data   **Future dates of conjunction**

3 Dec 1999
4 Dec 2000
7 Dec 2001
9 Dec 2002
12 Dec 2003
13 Dec 2004
16 Dec 2005
18 Dec 2006
21 Dec 2007
22 Dec 2008
24 Dec 2009
27 Dec 2010

Future dates of opposition

1 Jun 2000
4 Jun 2001
7 Jun 2002
9 Jun 2003
11 Jun 2004
14 Jun 2005
17 Jun 2006
19 Jun 2007
21 Jun 2008
23 Jun 2009
25 Jun 2010

Recent data

To be perfectly honest, there is not much to report. At the time of writing, Pluto remains the one planet not visited by human spacecraft. Until that time, ground-based observations, coupled with those from orbiting observatories, will be depended upon to provide all data on Pluto.

Late-1990s observations by the Hubble Space Telescope show that Charon is bluer than Pluto. Planetary scientists interpret this as showing that these two bodies have different surface composition and structure. A bright highlight on Pluto indicates that it might have a smoothly reflecting surface layer. A detailed analysis of the image suggests that there is a bright area parallel to the equator of Pluto. Subsequent observations are needed to confirm that this feature is real. The image was taken when Charon was near its maximum elongation from Pluto, which, at the distance of the Earth, amounts to only 0.9 seconds of arc.

The images from the Hubble Space Telescope show that Pluto is an unusually complex object, with more large-scale contrast than any planet except Earth. The resolving power of the Space Telescope is approximately 161 km per pixel. At this resolution, Hubble discerns roughly 12 major "regions" where the surface is either bright or dark.

Because of these images, Space Telescope scientists were able to construct the first map of the surface of Pluto. The map, which covers 85% of the planet's surface, confirms that Pluto has a dark equatorial belt and bright polar caps. This is in agreement with ground-based light curves obtained during the mutual eclipses that occurred between Pluto and Charon in the late 1980s.

Most of Pluto's surface features are likely to be produced by the complex distribution of frosts that migrate across Pluto's surface due to its orbital and seasonal cycles. This is easy to imagine since, over the course of its orbit, Pluto's distance from the Sun changes by more than 3.1 billion km! Additional surface features may be chemical byproducts deposited from Pluto's nitrogen–methane atmosphere. Names may eventually be proposed for some of the larger regions.

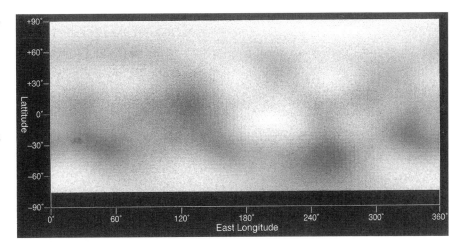

The surface of Pluto, as imaged by the Hubble Space Telescope. Courtesy of A. Stern (Southwest Research Institute), M. Buie (Lowell Observatory), NASA and ESA.

Satellite

	Size (km)	Mass (kg)[a]	Density (g/cm³)	Orbital period	Eccentricity	Inclination	Distance from planet (km)
Charon	1172	1.7×10^{21}	1.800	$06^{d}09^{h}17.3^{m}$	0.00	96.56°	1.9405×10^{4}

Historical timeline

6 April 1929	US observing assistant Clyde William Tombaugh took the first plate through the 33-cm A. Lawrence Lowell Astrographic refractor. Thus began his search for Planet X.
23 January 1930	The first of the discovery plates of Pluto was taken by Clyde Tombaugh.
29 January 1930	The second of the discovery plates of Pluto was taken.
18 February 1930	At 4 p.m. local time, Clyde Tombaugh discovered the motion of Pluto using a blink comparator.
13 March 1930	The discovery of Pluto was announced.
1954	Pluto's rotation rate was first determined.
1976	Spectroscopic analysis revealed the presence of methane on Pluto.
22 June 1978	US astronomer James Christy discovered Charon.
13 June 1983	Pioneer 10 crossed the orbit boundary of Pluto.
7 May 1990	Pluto's most recent closest approach to Earth.
11 February 1999	At 23:22 UT, Pluto crossed the orbit of Neptune and once again assumed the position of the most distant planet from the Sun. It will maintain that status until the year 2227.
2115	Pluto will be at maximum opposition distance from the Earth.

Appendix A
Short biographies of
some astronomers

Adams, John Couch (1819–1892)   British mathematician and astronomer who, along with LeVerrier, correctly predicted the existence and position of Neptune. Adams also refined the theory of the Moon's motion and did pioneering work on meteor showers.

Antoniadi, Eugène Marie (1870–1944)   Turkish-born French astronomer and planetary observer who is probably best remembered for his numerous observations of Mercury, mainly during daylight hours to improve the contrast of the planet against the background sky. Antoniadi published *La Planète Mercure* in 1934. He also expanded Schiaparelli's system of Martian nomenclature and published several maps of Mars.

Aristotle (384–322 BC)   Greek philosopher who wrote about all aspects of astronomy. Aristotle placed an immobile Earth at the center of all things. To account for observed irregularities, Aristotle theorized a system of rotating, nested spheres upon the surfaces of which rested the planets. This was an incredibly complex system, requiring 55 spheres, 22 of which were counterrotating.

Barnard, Edward Emerson (1857–1923)   US astronomer who was one of the great visual observers of all time and a pioneer of astrophotography. On 9 September 1892, Barnard discovered the fifth satellite of Jupiter, later named Amalthea. This was the last visual discovery of a planetary satellite. Barnard's observations of Saturn helped to clarify the nature of that planet's rings.

Beg, Ulugh (1394–1449)   Persian astronomer who built an observatory in Samarkand which contained a sextant 18.3 m in radius. His principal contribution to planetary astronomy was his set of accurate observations of the planet Venus. He also produced the Alfonsine Tables.

Bessel, Friedrich Wilhelm (1784–1846)   German astronomer best known for his positional determinations of stars. Bessel reserved some of his time for observations of the planets, notably Mercury and Saturn. He was the first to determine the parallax of a star, 61 Cygni. Bessel was the first Director of the Königsberg Observatory, a post he held for the last 36 years of his life.

Brahe, Tycho (1546–1601)   Danish astronomer regarded as the greatest of the pre-telescopic observers, Tycho made numerous planetary observations. Whereas earlier astronomers had been content to observe the positions of planets and the Moon at certain important points of their orbits (e.g. opposition, conjunction), Tycho and his cast of assistants observed these bodies throughout their orbits. As a result, a number of orbital anomalies never before noticed were discovered by Tycho. Without this complete series of observations of unprecedented accuracy, Kepler could not have discovered that planets move in elliptical orbits. Although he made tremendous contributions to planetary astronomy, Tycho did not accept the theory of Copernicus. He kept the Earth at the center of the solar system. In his arrangement, the Sun and Moon revolved around the Earth, but the planets revolved around the Sun.

Burroughs, Edgar Rice (1875–1950)   US novelist who wrote a series of fantastic adventures based upon his notion of life on Mars.

Cassini, Jean-Dominique (Giovanni Domenico) (1625–1712)   French astronomer and one of the greatest of the planetary observers, Cassini was the first Director of the Paris Observatory. He developed a method of mapping successive phases of solar eclipses, published new tables of the Sun and formulated a new theory of cometary orbits. He obtained powerful new telescopes and embarked on detailed studies that resulted in important findings about Mars (markings on the surface and the planet's period) and Jupiter (the revolution of the satellites, the rotation of the planet, spots, bands and flattening at the poles). He discovered four moons of Saturn: Iapetus, Rhea, Tethys and Dione. In 1675, he discovered that Saturn's rings are split largely into two parts by a narrow gap – now known as the "Cassini Division."

Christy, James (b. 1938)   US astronomer who, while working for the US Naval Observatory in 1978, was studying photographic plates of Pluto (working on improving Pluto's orbit parameters) when he noticed that Pluto had an irregular blob attached to its side which seemed to move around Pluto every 6.4 days. Charon was thus discovered.

Common, Andrew Ainslee (1841–1903)   British astronomer and pioneer of astrophotography, Common took the first successful photograph of Saturn's rings in 1883.

Copernicus, Nicolas (1473–1543)   Polish astronomer often referred to as the father of modern astronomy. His heliocentric theory, published in *De Revolutionibus Orbium Coelestium* (1543), placed the Sun in the center of the solar system, with the Earth and the other planets revolving around it. His theory was not perfect, however. Copernicus kept the idea of circular orbits, an idea which would fall with the planetary laws of Kepler.

Dawes, William Rutter (1799–1868)   British astronomer who was a tireless observer and illustrator of the planets, especially Mars and Saturn.

Democritus (*c.* 460–370 BC)   Greek philosopher who claimed that the Moon was similar to the Earth. Democritus also developed a world theory in which all matter was made of small particles called atoms.

Denning, William Frederick (1848–1931)   British amateur astronomer and author, Denning made numerous observations of the planets in order to determine their rotational rates. While he regularly observed Mercury, Venus, Mars and Saturn, the majority of his attention was directed toward Jupiter. Denning made a detailed historical study of the Great Red Spot's appearance and recorded many thousands of Jovian satellite transits.

Encke, Johann Franz (1791–1865)   German astronomer who did detailed calculations of the orbits of comets and asteroids. Encke observed a dark band in the A-ring of Saturn. This would later be known as the Encke Division.

Eudoxus of Knidus (*c.* 400–347 BC)   Greek philosopher who, like Aristotle, utilized a series of geocentric, rotating spheres to account for irregularities in planetary motion.

Galilei, Galileo (1564–1642)   Italian astronomer and inventor, Galileo did not invent the telescope. He did, however, improve on that invention (by the Dutch spectacle-maker Hans Lippershey) and made a number of "optic tubes." With his telescopes, Galileo made many discoveries, including craters on the Moon, sunspots, the phases of Venus and the four large moons of Jupiter. This last discovery was observational proof against the geocentric theories of his time, and caused Galileo no small amount of suffering.

Gassendi, Pierre (1592–1655)   French astronomer who, on 7 November 1631, became the first to witness a transit of Mercury. This observation confirmed Kepler's prediction and, indirectly, the heliocentric theory of Copernicus.

Hall, Asaph (1829–1907)   US astronomer who, while working at the US Naval

Observatory in 1877, discovered the two satellites of Mars, which he named Phobos and Deimos. Hall also made numerous observations of planetary satellites, asteroids and double stars.

Halley, Edmond (1656–1742) British astronomer who made numerous discoveries. From 1676 to 1678, from the island of Saint Helena, Halley catalogued the positions of about 350 Southern Hemisphere stars and observed a transit of Mercury. He proposed using transits of Mercury and Venus to determine the solar parallax. Using his theory of cometary orbits he calculated that the comet of 1682 was periodic, and predicted that it would return in 76 years. It did and it is now known as Halley's Comet. Halley was also the man most responsible for getting Newton's theory of gravitation published.

Harkness, William (1837–1903) US astronomer best known for his improvements in telescope design. Harkness also studied the planets and, in 1889, measured the mass of Mercury.

Heraclides Ponticus (388–315 BC) Greek philosopher born in what is now Turkey. Heraclides proposed that Mercury and Venus revolve around the Sun, and that the Sun revolves around the Earth. He was also the earliest known philosopher to state that the Earth rotates on its axis once a day.

Herschel, Sir William (1738–1822) German-born British astronomer who discovered the planet Uranus in 1781. He also discovered two satellites in orbit around both Saturn and Uranus. Herschel also constructed a number of large reflecting telescopes, determined the direction of the solar apex and made voluminous contributions to the study of stellar astronomy by conducting star counts and cataloguing more than 2000 nebulae and 800 double stars.

Hevelius, Johannes (1611–1687) Polish astronomer who was noted for his maps of the Moon, which first appeared in his *Selenographia*, in 1647. Many of the names given to lunar features by Hevelius are still used today. He also predicted and observed a transit of Mercury in 1661.

Hipparchus (second century BC) Turkish-born Greek philosopher who is best known for his discovery of precession. Hipparchus, using a lunar eclipse, calculated a reasonably accurate distance to the Moon. And although he maintained circular motion, he was the first to offset the orbit of the Earth to account for seasonal variations.

Hooke, Robert (1635–1703) British physicist and inventor, Hooke was probably the first to see Jupiter's rotation and the planet's Great Red Spot, in 1664.

Horrocks, Jeremiah (1618–1641) British astronomer who studied many aspects of planetary astronomy. Horrocks improved the theory of the Moon's motion and began a study of the tides. In 1639, he – along with his friend William Crabtree – became the first to observe a transit of Venus.

Huygens, Christiaan (1629–1695) Dutch astronomer and optician, Huygens built many refracting telescopes during his lifetime. He invented the pendulum clock. Huygens discovered the satellite Titan and was the first to correctly identify the nature of Saturn's rings.

Keeler, James Edward (1857–1900) US astronomer who was the first to clearly see Encke's Division. Keeler also used spectroscopic analysis to determine the meteoric nature of Saturn's rings.

Kepler, Johannes (1571–1630) German astronomer and theoretician, he is best remembered for his three Laws of Planetary Motion, a full discussion of which would more than fill this book. In addition, Kepler dramatically improved the science of celestial mechanics and positional astronomy.

Kirkwood, Daniel (1814–1895) US astronomer primarily known for his studies of the smaller members of the solar system. Kirkwood discovered that the gravity of Jupiter causes gaps in the asteroid belt. He also found a number of resonances between the gaps in Saturn's rings and the satellites of that planet.

Kuiper, Gerard Peter (1905–1973)   Dutch astronomer who discovered the satellites Miranda and Nereid. Kuiper refined the analysis of planetary atmospheres and was the first to detect methane on Titan.

Langley, Samuel Pierpont (1834–1906)   US astronomer and founder of the Smithsonian Astrophysical Observatory, Langley was the first to see Mercury silhouetted against the solar corona.

Laplace, Pierre Simon, marquis de (1749–1827)   French theoretician who made many contributions to the science of celestial mechanics and who added much to planetary motion theories. Laplace suggested that Saturn was circled by a large number of (solid) rings.

Lassell, William (1799–1880)   British astronomer who discovered the satellites Ariel, Umbriel and Triton.

LeVerrier, Urbain Jean Joseph (1811–1877)   French astronomer who, along with John Couch Adams, correctly predicted the existence and position of Neptune. LeVerrier's request for a search was acted upon quickly, thus the German astronomer J. G. Galle was the first to see the planet.

Lomonosov, Mikhail Vasilievich (1711–1765)   Russian physicist who, during the transit of 1761, discovered that Venus has an atmosphere.

Lowell, Percival (1855–1916)   US astronomer who built an observatory in Flagstaff, AZ, specifically for the study of the planet Mars. Lowell firmly believed that Mars was inhabited by intelligent life and wrote extensively to promote his views.

Maraldi, Giacomo Filippo (1665–1729)   Italian astronomer who made extensive planetary observations. In 1704, he first saw white spots at the poles of Mars. Fifteen years later, he suggested that the spots were ice caps.

Marius (Mayr), Simon (1573–1624)   German astronomer who attempted to claim precedence in the discovery of the four large moons of Jupiter. It was Marius who dubbed them "satellites." He also gave them the names by which they are known today.

Newcomb, Simon (1835–1909)   Canadian-born US astronomer who contributed greatly to planetary science. Newcomb precisely calculated the mass of Jupiter, confirmed the advance of Mercury's perihelion and refined the theory of the Moon's motion to a high accuracy.

Newton, Sir Isaac (1642–1727)   British mathematician and philosopher who derived the theory of universal gravitation, published in 1687 in his *Philosophiae naturalis principia mathematica*. He also invented calculus.

Pickering, William Henry (1858–1938)   US astronomer who discovered the satellite Phoebe. Prior to the discovery of Pluto, Pickering did a great deal of theoretical research into the existence of trans-Neptunian planets, none of which were ever found. He was also a pioneer of celestial photography.

Ptolemy, Claudius (c. 100–170)   Greek theoretician whose major work, the *Almagest* (*Syntaxis*), formed the basis for scientific thought for 15 centuries. Unfortunately, many of his conclusions were incorrect. For example, Ptolemy taught that all heavy matter tends to move to the center of the world, thus, if the Earth rotated, everything would be torn apart.

Roche, Edouard (1820–1883)   French astronomer who proposed the most widely accepted theory relating to the formation of Saturn's rings, that a satellite had approached Saturn so closely that tidal forces had broken it apart. Today, astronomers define "Roche's limit" as the minimum distance from the center of a planet that a satellite can be without being destroyed by tidal forces.

Schiaparelli, Giovanni Virginio (1835–1910)   Italian astronomer who developed a system of naming features on the Martian surface. Along with Father Secchi, Schiaparelli made use of the Italian term "canali." He also made the first detailed map of Mercury and concluded that Mercury was tidally locked to the Sun. He imagined a 1:1 lock; however, the actual ratio was found to be approximately 3:2.

Schroeter, Johann Hieronymus (1745–1816)   German astronomer who constructed an observatory at Lilienthal, from which he became the first to systematically observe the surface of the Moon and the planets over a long period of time. Schroeter also discovered and named a number of lunar features.

Secchi, Pietro Angelo (1818–1878)   Italian astronomer and Jesuit monk, also known as Father Secchi, who drew the first color sketches of Mars. Secchi was the first to coin the descriptive term "canali," the Italian word for "channels" to describe linear features that he saw on Mars.

Slipher, Vesto Melvin (1875–1969)   US astronomer who refined the rotational periods of the planets and did pioneering work in the study of planetary atmospheres.

Tombaugh, Clyde William (1906–1997)   US astronomer who, in 1930, while working at the Lowell Observatory in Flagstaff, AZ, discovered Pluto. After finding Pluto, Tombaugh spent an additional 13 years examining over 90 million additional star images. However, he discovered no other planets.

Trouvelot, Leopold (1827–1895)   French astronomer who made (unconfirmed) sightings of surface features on Mercury.

Ussher, Archbishop James (1581–1656)   British theologian who fixed the date of creation as 22 October 4004 BC. He based this calculation on the descendants of Adam, as listed in the book of Genesis.

Appendix B
Unit conversion table

1 astronomical unit (AU)	=	149 597 870 kilometers
Celsius (°C)*	=	Kelvin −273°
		(To change Celsius to Farhenheit, multiply by 9/5 and add 32)
1 centimeter (cm)	=	0.3937 inch
1 cubic centimeter (cm³)	=	0.003 531 foot³
	=	0.061 02 inch³
1 cubic kilometer (km³)	=	0.239 91 mile³
declination is measured in degrees north or south of the celestial equator		
1 gram (g)	=	0.035 27 ounce
	=	0.002 2046 pounds
Kelvin (K)*	=	Celsius +273
1 kilogram (kg)	=	0.001 1023 ton
	=	2.2046 pounds
1 kilometer (km)	=	0.621 37 mile
1 kilometer per second (km/s)	=	3280.8 feet/second
1 kilometer per hour (km/hr)	=	0.911 34 feet/second
1 magnitude	=	a brightness difference of 2.511 8865
5 magnitudes	=	a brightness difference of 100
1 meter (m)	=	3.2808 feet
	=	39.37 inches
	=	1.0936 yards
1 meter/second (m/s)	=	2.2369 miles/hour
1 meter per second squared (m/s²)	=	details to come
1 pascal	=	0.000 009 869 atmosphere
right ascension is measured in hours, minutes and seconds eastward from the vernal equinox		
speed of light in a vacuum	=	299 792 458 meters/second
	=	186 282 miles/second
1 tesla	=	1 weber/meter²
1 watt (W)	=	4.1868 calories per second

* Temperature in degrees Centigrade is temperature in Kelvin minus 273.

Glossary

accretion
the accumulation of dust and gas into larger bodies.

albedo
the reflectance of a planet, satellite or other non-luminous object; the ratio of the total amount of light reflected in all directions. A perfect reflector would have an albedo of 1.0; a black surface which absorbs all light would have an albedo of 0.0. Albedo may be divided into two types: Bond albedo and geometric albedo. All albedo values used in this book are visual geometric albedo.

albedo feature
a dark or light marking on the surface of an object that might not be a geological or topographical feature.

altitude
the angular distance of a point or celestial object above or below the horizon. It is measured along the vertical circle through the body from 0° (on the horizon) to 90° (at the zenith). Negative values correspond to objects which lie below the horizon.

angular size
the apparent size of a celestial object, measured in degrees, minutes and seconds, as seen from the Earth; for example, the average angular size of the Sun, as seen from Earth, is 0.53°.

antipodal point
the point that is directly on the opposite side of the planet; e.g., the Earth's north pole is antipodal to its south pole.

aphelion
the position of an object in solar orbit when it is furthest from the Sun; the instant in a given orbit of a planet (or other body) when it is furthest from the Sun.

apoapsis
the position of an object in orbit about a planet that is furthest from the planet.

apogee
the position of the Moon or another object in Earth orbit when it is furthest from the Earth; the instant in a given orbit of the Moon (or other object) when it is furthest from the Earth.

apojove
the position of an object in orbit around Jupiter when it is furthest from the planet; the instant in a given orbit of a satellite (or other body) when it is furthest from Jupiter.

apomartian
the position of an object in orbit around Mars when it is furthest from the planet; the instant in a given orbit of a satellite (or other body) when it is furthest from Mars.

aposaturnian
the position of an object in orbit around Saturn when it is furthest from the planet; the instant in a given orbit of a satellite (or other body) when it is furthest from Saturn.

apparent magnitude
see magnitude.

apparition
the period of time during which a celestial body may be observed.

aquifer	body of rock sufficiently permeable to conduct ground fluid such as water on Earth or sulfur dioxide on Io.
Arean	of or pertaining to Mars; more commonly *Martian*.
ascending node	*see* nodes.
aspect	the position of the Moon or planets relative to the Sun, as seen from Earth. Conjunctions, oppositions and quadratures are examples of aspects. Also known as *configuration*.
asteroids	small bodies composed of rock and metal which orbit the Sun. Most (95%) lie in a belt between the orbits of the planets Mars and Jupiter. Also known as *minor planet*.
astronomical unit	a unit of distance which is approximately (within about 3/100 000 000) the average distance from the Earth to the Sun; this distance is approximately 149 597 870 km. Abbreviated AU.
atmosphere	the gaseous layer which surrounds a celestial body.
atmospheric pressure	the pressure exerted by the atmosphere of a celestial body at any given location within that atmosphere; as a unit of measurement, one atmosphere is the pressure the Earth's atmosphere exerts at sea level, approximately 101 325 newtons per square meter (101 325 pascals) or 14.7 pounds per square inch or 1013 millibars; a pressure of one atmosphere is sometimes referred to as a *bar*, to which it is approximately equal (1 bar = 0.987 atmosphere).
AU	*see* astronomical unit.
aurora	a glow in a planet's ionosphere caused by the interaction between the planet's magnetic field and charged particles from the Sun (the solar wind).
axis	the imaginary line around which a rotating body (such as a planet or satellite) turns.
azimuth	the angular distance to an object measured eastward along the horizon from the north to the intersection of the object's vertical circle; varies from 0° to 360°. Thus, an object due east would have an azimuth of 90°, and an object due west would have an azimuth of 270°.
bar	a unit of atmospheric pressure; 1 bar = 0.987 atmosphere = 100 008 pascals.
basalt	a general term for dark-colored, igneous rocks composed of minerals that are relatively rich in iron and magnesium.
Bond albedo	the fraction of incident light that is reflected from a solar system object.
bow shock	a shock front set up between a planetary magnetic field and the solar magnetic field due to the orbital motion of the planet.
breccia	a coarse-grained rock, composed of angular, broken rock fragments held together by a mineral cement or a fine-grained matrix.

brightness — *see* magnitude.

brilliancy at opposition — the apparent magnitude of a celestial body when at opposition, 180° from the Sun, as seen from the Earth.

butte, buttes — a conspicuous, isolated, flattop hill with steep slopes.

caldera, calderas — a large, basin-shaped volcanic depression, more or less circular, the diameter of which is many times greater than that of the included vent or vents.

carbonate — a compound containing carbon and oxygen.

Cassini Division — a relatively empty area of Saturn's ring system, 4200 km wide, located between the A- and B-rings.

catena, catenae — a chain of craters on the surface of a planet or a satellite.

cavus, cavi — hollow; irregular depression on the surface of a planet or a satellite.

celestial equator — the intersection of the equatorial plane of the Earth with the celestial sphere; the projection of the Earth's equator onto the sky.

celestial sphere — the apparent background of the stars, assumed to be of infinite extent in all directions; the sky.

central peak — central high area produced in an impact crater by inward and upward movement of underlying material.

central pit — ring formation produced within an impact crater, analogous to a central peak.

chaos — a distinctive area of broken terrain on the surface of a planet or a satellite.

chasma, chasmata — a canyon on the surface of a planet or a satellite.

cloud features — visible features in the atmosphere of a planet or satellite; may be temporary or permanent.

colles — a small hill or knob on the surface of a planet or a satellite.

comet — one of many relatively small solar system objects which orbit the Sun; comets are composed of frozen gases and dust; the parts of a comet include the nucleus, which is the comet itself; a head may develop when the comet is near the Sun; one or more tails may also develop, composed of dust or gas; short-period comets have orbital periods which are less than 200 years; long-period comets have periods greater than 200 years.

conjunction — the alignment of two celestial objects such that the difference in their longitude, as seen from Earth, is 0°. Two objects may also be in conjunction in right ascension. When one of the objects is the Sun "conjunction" denotes when the other object is in line with the Sun, and therefore usually invisible.

constellation	one of 88 arbitrary configurations of stars; the area of the celestial sphere containing one of these configurations.
Copernican system	a model of our solar system introduced by Nicolas Copernicus. It was published in 1543, in the book *De Revolutionibus Orbium Coelestium (On the Revolutions of the Celestial Spheres)*. It placed the Sun in the center of the solar system, rather than the Earth.
Coriolis Force	the force which causes a deflection, with respect to the ground, of an object moving above the surface of the rotating Earth; the Coriolis Force most notably affects the movement of large storm systems.
corona, coronae	an ovoid-shaped feature found only on the surfaces of Venus and Miranda.
craters	roughly circular depressions on the surface of many objects in the solar system; most craters were created through meteoritic impact, with the remainder being volcanic or caused by surface collapse.
crescent	a phase of the Moon or other celestial body where the percent of visible surface illumination is greater than 0% but less than 25% (or 50% of the side currently being observed).
culmination	the passage of a celestial body across an observer's meridian. "Upper culmination" (also called "transit") is the crossing nearer to the observer's zenith; "lower culmination" is the crossing further from the observer's zenith.
Cytherean	of or pertaining to Venus; more commonly *Venerian*.
day	generally defined as one rotation of a planet on its axis. A sidereal day is measured with respect to the stars, while a solar day is measured with respect to the Sun. Many other types of days are defined.
dayglow	the resonant absorption and emission of sunlight at a particular wavelength in the atmosphere of a planet or satellite.
declination	an Earth-centered angle measured perpendicularly from the celestial equator to a point on the celestial sphere. Declination is positive if the object or point is north of the celestial equator and negative if the object or point is south of the celestial equator.
degree (of arc)	1/360 of a circle; one degree of arc contains 60 minutes of arc; designated °; thus 85°18'08" which is 85 degrees, 18 minutes, 8 seconds.
density	with respect to a planet, satellite or other celestial object, the density equals the mass of the object divided by its volume, usually measured in grams per cubic centimeter.
descending node	*see* nodes.
diameter	the length from the surface of a celestial object, through its center, to the surface on the other side; the diameter is twice the radius.
dichotomy	the moment when the Moon, Mercury or Venus appears exactly half illuminated.

differentiation	processes by which planets and satellites develop concentric layers or zones of different chemical and mineral composition.
direct motion	the apparent (west to east) motion of a planet or other celestial object on the celestial sphere, as seen from Earth.
disk	the visible surface of any heavenly body projected against the sky.
dorsum, dorsa	a ridge on the surface of a planet or a satellite.
eccentricity	a measurement (from 0 to 1) which is the amount that the orbit of any solar system object is not circular. An object in a circular orbit would have an eccentricity of 0. Mathematically, this is defined as the distance between the focal points of an ellipse divided by twice the length of the major axis.
eclipse	the obscuration of light from a celestial body as it passes through the shadow of another body; such obscuration may be total or partial.
ecliptic	the great circle described by the Sun's annual path on the celestial sphere.
ecliptic plane	the mean plane of the Earth's orbit around the Sun.
ejecta blanket	the deposit surrounding an impact crater composed of material ejected from the crater during its formation.
electroglow	a phenomenon in the atmosphere of Uranus causing it to glow brightly in ultraviolet light.
ellipse	the shape of all orbits in the solar system. Mathematically, a type of conic section with an eccentricity less than one. A circle is defined as an ellipse having an eccentricity of zero.
elongation	the apparent angle subtended by the Sun and a planet or by a planet and one of its satellites as seen from Earth. It is measured from 0° to 180° east or west of the Sun and from 0° east or west of the planet.
Encke Division	a division in the Saturnian ring system which separates the A-ring and the F-ring; the Encke Division is approximately 325 km in width.
endogenic	a geologic process that derives its energy from the interior of a planet or satellite, such as a volcanic eruption.
eolian	related to wind deposits and associated effects.
Ephemeris Time (ET)	the time scale that together with the laws of motion correctly predicts the positions of celestial bodies, and it is therefore used as the argument in the ephemerides. The current Ephemeris Time is thus determined by comparing the observed positions with the ephemerides.
equator	the great circle on the surface of a rotating celestial body that lies in the plane that passes through the center of the body and is perpendicular to its axis of rotation.

equilibrium (in cratering)	condition of constant crater population in which new impacts or other processes destroy old craters as rapidly as new ones are created.
equinox	either of two points on the ecliptic, lying at right ascension 0 hours (March equinox) and 12 hours (September equinox).
eruption	the ejection of volcanic materials (lavas, pyroclasts and volcanic gases) onto the surface of a planet or satellite, from a central vent, a fissure or a group of fissures.
eruptive center	an active volcanic center, usually referring to those found on Io.
escape velocity	the minimum velocity required for an object to escape the gravitational attraction of a planet or other celestial body. If the object fails to attain escape velocity, it will enter into an elliptical orbit around the planet; generally, the escape velocity is calculated from the equator of a celestial body and is thus the equatorial escape velocity.
evening star	a term often used to describe the appearance of a bright planet (Mercury, Venus, Mars, Jupiter or Saturn) in the western evening sky.
exogenic	a geological process that is caused by an external force, such as a meteorite impact.
facula, faculae	a bright spot on the surface of a planet or a satellite.
farrum, farra	a pancake-like structure, or a row of such structures, on the surface of a planet or a satellite.
fault	a fracture or zone of fractures along which the sides are displaced relative to one another, parallel to the fracture.
First Contact	during an eclipse, the moment that the shadow of the eclipsing body first makes contact with the body being eclipsed; the beginning of the eclipse.
fissure	a narrow opening or crack of considerable length and depth on the surface of a planet or satellite.
flexus	a cuspate linear feature on the surface of a planet or a satellite; a very low curvilinear ridge with a scalloped pattern.
fluctus	a flow terrain on the surface of a planet or a satellite.
fossa, fossae	a long, narrow, shallow depression on the surface of a planet or a satellite.
Fourth Contact	during an eclipse, the moment that the shadow of the eclipsing body breaks contact with the body being eclipsed; the end of the eclipse.
full	a phase of the Moon or other celestial body where the percent of visible surface illumination is 50% (or 100% of the side currently being observed).
fumarole	volcanic vent from which gases are emitted.

Galilean satellites	the four large satellites of Jupiter (Io, Europa, Ganymede and Callisto) discovered in 1610 by Galileo.
gauss	*see* tesla.
geometric albedo	how bright a solar system object is relative to a sphere of equal size made of perfectly reflecting diffuse white material.
geyser	a miniature volcano from which hot water and steam erupt periodically.
gibbous	a phase of the Moon or other celestial body where the percent of visible surface illumination is greater than 25% but less than 50% (or greater than 50% but less than 100% of the side currently being observed).
graben	an elongated, relatively depressed crustal unit or block that is bounded by faults on its sides.
gravity	the attractive force of all bodies possessing mass.
Great Red Spot	an oval-shaped storm in the atmosphere of Jupiter, located 22° south of Jupiter's equator.
greatest brilliancy	the points in the orbits of Mercury and Venus when they appear brightest as seen from Earth.
greenhouse effect	the heating of a planetary atmosphere through the following process: Visible radiation from the Sun passes through the atmosphere and is absorbed by the surface. It is later re-radiated as infrared radiation, which is absorbed and re-emitted by the atmosphere.
Gregorian calendar	the calendar in general use today, introduced in 1582 by Pope Gregory XIII as a revision of the Julian calendar, adopted in Great Britain and the American colonies in 1752, and having leap years in every year divisible by four with the restriction that centennial years are leap years only when divisible by 400; for example, the years 1700, 1800 and 1900 are not leap years, but 2000 is.
Hadean	of or pertaining to Pluto; *more commonly Plutonian.*
horizon	where the celestial sphere intersects the Earth at every point; where the sky meets the Earth.
inclination	the angle between the orbital plane of a planet and the ecliptic plane.
inferior conjunction	the position of an inferior planet (Mercury or Venus) when it is in conjunction with the Sun and between the Sun and Earth.
inferior planet	any planet whose orbit around the Sun is closer than that of the Earth; Mercury or Venus.
infrared radiation	a type of electromagnetic radiation with wavelengths longer than those of visible light, but shorter than those of radio waves.

inner planets	planets closer to the Sun than the asteroid belt; Mercury, Venus, Earth and Mars.
insolation	direct solar and sky radiation reaching a body; also, the rate at which it is received, per surface unit.
interplanetary	a term which denotes the space between the planets.
interplanetary medium	the material contained in the solar system in the space between the planets. One of the main components of this space is the solar wind.
invariable plane	the plane through the center of mass of the solar system; the plane which represents the mean of the orbital planes of all the planets; at the time of writing the invariable plane is inclined 1° 39′ to the plane of the ecliptic.
ionosphere	a region of charged particles in a planet's upper atmosphere; the part of the Earth's atmosphere beginning at an altitude of approximately 400 km and extending outward 400 km or more.
irradiance (solar)	the amount of radiant energy per unit area reaching a planet or other solar system body.
Jeans escape	the escape of gas from a planetary atmosphere caused by the thermal motion of the gas molecules.
Jovian	of or pertaining to Jupiter.
Jovian planets	the four large outer planets; Jupiter, Saturn, Uranus and Neptune; named due to the similarities in size and composition with Jupiter.
Julian calendar	a calendar introduced in Rome in 46 BC establishing the 12-month year of 365 days with each fourth year having 366 days and the months each having 31 or 30 days except for February which has 28 or in leap years 29 days; *also see* Gregorian calendar.
Keeler Division	a division 35 km in width, lying in the outer A-ring of the Saturnian ring system.
kelvin	a unit of temperature for the absolute temperature scale. At absolute zero – 0 K – molecules of all substances have no heat energy. The temperature in K is that in °C plus 273.15.
Kepler's laws	the three basic tenets of planetary motion, introduced by Johannes Kepler. (1) Each planet moves in an elliptical orbit with the Sun at one focus of the ellipse. (2) A line joining the planet and the Sun sweeps out equal areas in equal amounts of time. This indicates that a planet travels fastest when closest to the Sun and slower when further away. (3) The square of the period of a planet's orbit is proportional to the cube of the distance of the planet from the Sun. This indicates that any planet closer to the Sun than another planet would move faster in its orbit and also allowed a relative distance scale for the solar system to be established.
kilopascal	*see* pascal.

labes	a landslide on the surface of a planet or a satellite.
labyrinthus	an intersecting valley complex on the surface of a planet or satellite.
lacus	a lake-like structure on the surface of a planet or satellite; a small plain.
Lambert surface	a diffuse, perfect reflector at all wavelengths.
Laplace relation	a three-way orbital resonance involving Io, Europa and Ganymede.
lava	a general term for molten rock that is extruded onto the surface of a planet or satellite.
lenticula, lenticulae	small dark spots on Europa.
libration	a small angular change in the face that a synchronously rotating satellite presents toward the center of its orbit; the phenomenon whereby more than 50% of the Moon's surface is visible to a terrestrial observer.
light-time	the time that light (which moves at 299 792 km/s in a vacuum) takes to travel between two objects or points.
limb	the outer edge of the apparent disk of a celestial body.
linea, lineae	a dark or bright elongate marking on the surface of a planet or a satellite. It may be curved or straight.
lithosphere	the stiff upper layer of a planetary body, including the crust and part of the upper mantle.
lunar	of or pertaining to the Moon.
m_{vis}	the apparent visual magnitude of a celestial object, as seen from Earth.
maar	a low-relief, broad volcanic crater formed by multiple shallow explosive eruptions.
macula, maculae	a dark spot on the surface of a planet or satellite.
magma	mobile or fluid rock material; lava; generalized to refer to any material that behaves like silicate magma in the Earth.
magnetic field strength	the strength of a planetary magnetic field measured at the surface of the planet, or at a given distance above the surface.
magnetopause	the boundary of a planet's magnetic field; where a planet's magnetic field begins to dominate the solar magnetic field.
magnetosheath	a turbulent magnetic layer lying between a planet's bow shock and its magnetopause.
magnetosphere	the region surrounding a planet where the planetary magnetic field is the controlling force for ionized particles, rather than the magnetic field of the Sun.

324

magnitude	within the confines of the solar system (for the purposes of this book) this usually refers to the *apparent visual magnitude*, which is the brightness of a celestial object as seen from Earth, irrespective of its true brightness.
major axis	the greatest distance across an ellipse; the distance from edge to edge through the center and both foci. For a circle, this distance would equal the diameter.
mare	Latin word for "sea." The term is still applied to the basalt-filled impact basins common on the face of the Moon visible from Earth.
Martian	of or pertaining to Mars.
mass	the amount of matter in an object; the property of a body that is a measure of its inertia and that is commonly taken as a measure of the amount of material it contains and causes it to have weight in a gravitational field.
Maxwell Division	a division, 270 km wide, lying between the B- and C-rings of the Saturnian ring system.
mensa, mensae	a mesa; flattopped elevation with cliff-like edges on the surface of a planet or a satellite.
Mercurial	of or pertaining to Mercury.
meridian	the great circle passing through the observer's zenith and the celestial poles.
mesa	a broad, flattop, erosional hill or mountain, commonly bounded by steep slopes.
meteor	a streak of light in the night sky caused by a meteoroid entering the Earth's upper atmosphere and burning due to friction with the atmosphere; sometimes called a "shooting star" or "falling star."
meteorite	any meteoroid which (after becoming a meteor) lands on the surface of the Earth.
meteoroid	small body made of rock, metal or a combination of both which orbits the Sun; most are extremely small, with masses between 1/1 000 and 1/1 000 000 of a gram.
Metonic cycle	a lunar cycle (discovered by the Greek astronomer Meton in the fifth century BC) equal to 19 years, after which the phases of the Moon will recur on the same days of the year.
minor axis	the distance from the edge of an ellipse, through the center, to the other edge, perpendicular to the line connecting the foci.
minor planet	*see* asteroid.
minute (of arc)	1/60 of one degree of arc. There are 60 seconds of arc in one minute of arc; designated '; thus 45°06'14" = 45 degrees, 6 minutes, 14 seconds.
mons, montes	a mountain on the surface of a planet or satellite.

morning star	a term often used to describe the appearance of a bright planet (Mercury, Venus, Mars, Jupiter or Saturn) in the eastern morning sky.
mutual event season	time when mutual phenomena of the Galilean satellites of Jupiter may be observed; these occur at approximately six-year intervals as the Earth crosses the orbital plane of the satellites.
mutual phenomena (of satellites)	the situation that occurs when one satellite – usually of Jupiter – occults or is eclipsed by a second satellite.
Neptunian	of or pertaining to Neptune.
new	a phase of the Moon or other celestial body where the percent of visible surface illumination is 0%.
nodes	the two points at which the orbital plane of a celestial body intersects a reference plane, such as the ecliptic or celestial equator. If the body is seen to move across the reference plane from south to north, the node is referred to as an ascending node. If the body is seen to move from north to south, the node is a descending node.
nutation	a periodic, irregular motion of the Earth caused by the gravitational attraction of the Sun and Moon, along with their varying distances and relative directions. This motion is superimposed upon the motion of precession.
oblateness	the measurement of the non-sphericity of a planet or other celestial object. Mathematically, this is the difference between the equatorial and polar diameters of a planet divided by the planet's equatorial diameter.
obliquity	the angle between the equatorial plane of a planet or other celestial object and its orbital plane; the angle between a planet's axis of rotation and the pole of its orbit.
occultation	the obscuration (total or partial) of any celestial object by another of larger apparent diameter.
oceanus	an ocean-like feature on the surface of the Moon.
Olympus Mons	the largest volcano on Mars. It has an elevation of 25 km with a base 550 km in diameter and a caldera 80 km in diameter.
opposition	the position of two celestial objects when their longitude (as seen from Earth) differs by 180°. When one of the objects is the Sun, opposition means the other object is opposite the Sun in the sky, therefore visible all night long.
opposition effect	the moment when the phase angle of a superior planet is 0°. This occurs when the Earth is in transit across the face of the Sun, as seen from the planet.
orbit–orbit resonances	a phenomenon in which two or more satellites interact gravitationally so that their motions follow certain repetitive patterns.
orbital elements	six parameters which specify the position and motion of a celestial body in its orbit and that can be established by observation. The six elements are eccentricity, semi-

major axis, inclination, longitude of the ascending node, argument of perihelion and epoch (sometimes referred to as time of perihelion passage).

orbital period — the period of time which a planet or satellite takes to revolve once about its primary, i.e., a year on that planet.

orbital velocity — the speed at which a planet orbits the Sun, or at which a satellite orbits a planet; this speed is greatest when the orbiting body is nearest the primary and least when the orbiting body is furthest from the object it is orbiting.

outer planets — planets further from the Sun than the asteroid belt; Jupiter, Saturn, Uranus, Neptune and Pluto.

palimpsest — a circular feature on the surface of dark icy moons such as Ganymede and Callisto lacking the relief associated with craters. Palimpsests are thought to be impact craters where the topographic relief of the crater has been eliminated by slow adjustment of the icy surface.

palus — a swamp-like feature on the surface of the Moon; a small plain.

parallax — the apparent angular displacement of a star or other celestial object that results from the revolution of the Earth about the Sun; for a star, numerically, this is the angle sub-tended by one astronomical unit at the distance of the particular object. This differs from the *solar parallax*, which is the apparent displacement of the Sun as seen from two (generally widely separated) places on Earth.

pascal — the standard unit of pressure; one pascal is defined as the pressure generated by a force of 1 newton acting on an area of 1 square meter; since this unit is quite small values are usually given in *kilopascals*; for conversion purposes, 1 atmosphere = 101 325 pascals.

patera, paterae — a crater with irregular or scalloped edges on the surface of a planet or a satellite; on Io, a volcanic vent surrounded by irregular flows.

penumbra — the less dark outer region of a shadow cast by a solar system object illuminated by the Sun.

periapsis — the position of an object in orbit about a planet that is closest to the planet.

perigee — the position of the Moon or an object in Earth orbit when it is closest to the Earth; the instant in a given orbit of the Moon (or other object) when it is closest to the Earth.

perihelion — the position of an object in solar orbit when it is closest to the Sun; the instant in a given orbit of a planet (or other body) when it is closest to the Sun.

perijove — the position of an object in orbit around Jupiter when it is closest to the planet; the instant in a given orbit of a satellite (or other body) when it is closest to Jupiter.

perimartian — the position of an object in orbit around Mars when it is closest to the planet; the instant in a given orbit of a satellite (or other body) when it is closest to Mars.

period	with regard to a celestial object, the time interval between two successive, similar events.
perisaturnian	the position of an object in orbit around Saturn when it is closest to the planet; the instant in a given orbit of a satellite (or other body) when it is closest to Saturn.
perturbation	the effect on the orbit of a planet or satellite as predicted by theory, produced by (1) another planet or satellite or (2) a group of planets and/or satellites.
phase	the percentage of illumination of the Moon or other solar system object at a particular time during its orbit.
photoionization	the removal of electrons from electrically neutral atoms or molecules resulting from the impact of high-energy electromagnetic radiation.
planet	one of nine solar system objects which orbit the Sun and shine by reflected light; a similar object in orbit around another star.
planetesimal	a hypothetical early body of intermediate size (1–100 m), usually finally accreted to a larger body.
planitia	a low plain on the surface of a planet or satellite.
planum	a plateau or high plain on the surface of a planet or satellite.
plume	a large pyroclastic and gas eruption observed on Io and on Triton.
Plutonian	of or pertaining to Pluto.
poles	two points on a planet, satellite or other celestial object which lie 90° above or below a given great circle, generally the object's equator.
precession	the sweeping out of a cone by the spin axis of a rotating body when acted upon by a torque perpendicular to its spin axis. The Sun and Moon (and to a much lesser extent, the planets) attract the equatorial bulge of the non-spherical Earth causing the poles to precess about a line through the Earth's center perpendicular to the ecliptic plane; thus, the celestial poles describe circles approximately 23.5° in radius on the celestial sphere.
primary	in a system of two or more orbiting celestial bodies, the one nearest to the center of mass; the one around which the others seem to revolve.
prograde	orbiting in the same direction as the prevailing direction of motion; the opposite of retrograde.
promontorium	a cape-like (headland) feature on the surface of the Moon.
Ptolemaic system	a model of our solar system completed by Claudius Ptolemaeus in the second century AD. It placed the Earth in the center of the solar system. All other objects revolved around the Earth, moving in circular motion. To correct for the observed irregularities

in this system, other circles of varying sizes were superimposed upon the larger orbital circles.

pyroclastic pertaining to material formed by a volcanic explosion or aerial expulsion from a volcanic vent.

quadrature the position of the Sun and another celestial object when their longitude (as seen from Earth) differs by 90°.

radiation belts regions of charged particles in a planetary magnetosphere.

radius the length from the center of a celestial object to its surface; the radius is half the diameter.

regio, regiones a large area distinguished by shading or color on the surface of a planet or satellite; a region.

regolith the layer of dust and broken rock, created by meteoritic bombardment, that covers much of the surface of the Moon and Mars.

residual the difference between the observed and the predicted (theoretical) positions of a planet.

reticulum, reticula reticular (net-like) pattern on Venus.

retrograde orbiting opposite to the prevailing direction of motion; the opposite of prograde.

retrograde motion the false (east to west) apparent motion of a planet or other celestial object on the celestial sphere, as seen from Earth. This occurs when a superior planet is near opposition. The Earth, moving faster, overtakes the planet and the planet appears to move backward. A similar phenomenon occurs when one automobile passes another even though both vehicles are traveling in the same direction, the slower automobile seems to be going the other way.

revolution the orbital motion of a planet or other celestial object around the Sun, or of a satellite around a planet.

rift a fracture or crack on the surface of a planet or satellite caused by extension.

rift valley an elongated valley formed by the depression of a block of the crust of a planet or satellite between two faults or groups of faults running approximately parallel.

right ascension a geocentric spherical coordinate that is an angle measured eastward along the celestial equator from the vernal equinox to the intersection of the hour circle passing through the body; usually expressed in hours, minutes and seconds from 0 hours to 24 hours, where one hour of right ascension equals 15°.

rima, rimae a fissure on the surface of a planet or satellite.

rising the appearance of a celestial object above the horizon due to the rotation of a planet or satellite.

Roche limit	the minimum distance from the center of a planet that a satellite can maintain equilibrium, that is, without being pulled apart by tidal forces. If a planet and a moon have the same density, the Roche limit is 2.446 times the radius of the planet.
rotation	the spinning motion of a planet or other celestial object around an axis.
rotational period	the period of time which a planet or satellite takes to spin once, measured at its equator; thus this is the equatorial period of rotation.
rotational velocity	the speed at which a planet or satellite spins; this is measured at the body's equator, thus this is the equatorial velocity of rotation.
rupes	scarps on the surface of a planet or satellite.
satellite	a body that revolves around a larger body.
saturation (in cratering)	the condition of maximum possible density of impact features on the surface of a planet or satellite.
Saturnian	of or pertaining to Saturn.
scarp	a line of cliffs produced by faulting or erosion; a cliff-like face or slope of considerable linear extent.
Schroeter effect	the fact that the dichotomy of Venus does not coincide with the time of greatest elongation. This was first described by the German amateur astronomer Johann Hieronymus Schroeter. This effect has also been noted for the Moon and Mercury.
scopulus	irregular scarps on the surface of a planet or satellite.
second (of arc)	1/60 of one minute of arc. Thus, there are 3600 seconds of arc in one degree of arc; designated ″; thus 22°31′46″ = 22 degrees, 31 minutes, 46 seconds.
Second Contact	during a total eclipse, the moment of total obscuration of the Sun or Moon; the instant totality begins.
secondary	in a system of two or more orbiting celestial bodies, any body not the closest to the center of mass which revolves around the primary.
semimajor axis	one-half the greatest distance across an ellipse; the distance from the center to the edge through one of the foci. For a circle, this distance would equal the radius.
semiminor axis	the distance from the center of an ellipse to the edge perpendicular to the line connecting the foci.
setting	the disappearance of a celestial object below the horizon due to the rotation of a planet or satellite.
shepherd satellite	a satellite that constrains the extent of a planetary ring through gravitational force.

shield volcano	a volcanic mountain in the shape of a broad, flattened dome.
sidereal period	an amount of time measured with reference to the stars.
silicate	a rock or mineral whose structure is dominated by bonds of silicon and oxygen atoms.
sinus	a bay or semienclosed break in a scarp on the surface of a planet or satellite.
size	the equatorial diameter of a planet or satellite; for a Jovian planet, size is measured from the limb of the planet, where the atmosphere becomes opaque.
solar	of or pertaining to the Sun.
solar irradiance	the amount of solar energy falling upon a unit area of the surface of a planet or satellite; usually measured in watts per square meter.
solar nebula	the gas and dust from which the solar system formed; the nebula surrounding the proto-Sun when the planets and smaller bodies were still accreting.
solar wind	energetic charged particles that flow radially outward from the solar corona, carrying mass and angular momentum away from the Sun.
solstice	either of two points on the ecliptic, lying at right ascension 6 hours (June solstice) and 18 hours (December solstice).
spectrum, absorption	the spectrum formed by a celestial object where wavelengths are manifested by patterns of dark lines or bands signifying a decrease in intensity of radiation at those wavelengths; such decreases allow the percentage of each of the absorbing substances to be measured, thus determining the composition of the object.
spectrum, continuous	a continuum of color formed when a beam of white light is dispersed so that its component wavelengths are arranged in order.
speed of light	the speed at which all electromagnetic radiation (not only light) propagates in a vacuum; the numerical value is 299 792 458 m/s.
stationary point	the position of a planet as it shifts from direct to retrograde motion, or from retrograde to direct, and thus appears motionless.
stratosphere	the cold region of a planetary atmosphere above the convecting regions, usually without vertical motion of the gases but sometimes exhibiting strong horizontal jet streams.
sulcus, sulci	an area of subparallel furrows and ridges on the surface of a planet or satellite.
superior conjunction	the position of an inferior planet when it is in conjunction with the Sun and on the side of the Sun opposite the Earth.
superior planet	any planet whose orbit around the Sun is further out than that of the Earth; Mars, Jupiter, Saturn, Uranus, Neptune and Pluto.

surface	for a terrestrial planet or a satellite, the boundary between the planet itself and its atmosphere; for a Jovian planet, this may refer to the boundary between the atmosphere and the solid core deep within or it may refer to the "optical" surface, beginning at a depth where the atmosphere becomes opaque.
surface gravity	the gravitational attraction of a body measured at its surface.
synchronous rotation	the condition arising when the rotational period of a satellite is equal to its orbital period.
synodic period	the average time between successive conjunctions of a planet as seen from the Earth.
syzygy	the lineup of the Sun, the Earth and either the Moon or a planet. For the Moon, this occurs at New Moon and Full Moon. For a planet, it occurs at conjunction and opposition.
telluric	of or pertaining to the Earth; more commonly *terrestrial*.
temperature	a measure of the heat energy given off by any celestial object.
terminator	the boundary between the sunlit and dark hemispheres of the Moon or other solar system object. At the terminator either sunrise or sunset is occurring.
terra	an upland area on the surface of a planet or satellite; an extensive land mass.
terrestrial	of or pertaining to the Earth.
terrestrial planets	the four small inner planets; Mercury, Venus, Earth, Mars; named due to the similarities in size and composition with Earth.
tesla	a unit of magnetic force equal to one weber per square meter; 1 tesla = 10 000 gauss.
tessera, tesserae	a tile-like area on the surface of a planet or satellite.
Third Contact	during a total eclipse, the moment that total obscuration of the Sun or Moon ends; the instant totality ends.
tholus, tholi	a hill or dome on the surface of a planet or satellite.
tidal force	a force on a planet, satellite or other solar system body caused by the differential gravitational attraction of one or more other bodies.
tidal heating	the frictional heating of a satellite's interior due to flexure caused by the gravitational pull of its parent planet and possibly neighboring satellites.
tilt of axis	the angle between the axis of a planet or satellite and its orbital plane.
transit	(1) the passage of Mercury or Venus across the Sun's disk, as seen from Earth (2) also known as *upper culmination*, this is the passage of a celestial body across an observer's meridian nearest to the zenith (to distinguish it from *lower culmination*).

troposphere	the lower regions of a planetary atmosphere where convection keeps the gases mixed and maintains a steady increase of temperature with depth.
twilight	the condition of solar illumination at a point on the Earth's surface before sunrise or after sunset when the Sun's zenith distance is 96° for civil twilight, 102° for nautical twilight or 108° for astronomical twilight.
ultraviolet radiation	a type of electromagnetic radiation with wavelengths shorter than those of visible light, but longer than those of x-rays.
umbra	the dark inner region of a shadow cast by a solar system object illuminated by the Sun.
uncompressed density	the density that an average fragment of a planet or satellite would have if it were not under high pressure.
undae	dune-like features on the surface of a planet or satellite.
Universal Time (UT)	the mean solar time of the prime meridian (which runs through Greenwich), determined by measuring the angular position of the Earth about its axis.
Uranian	of or pertaining to Uranus.
Valles Marineris	a vast system of Martian canyons extending for approximately 4128 km. The main canyon is about 7 km deep and up to 100 km wide.
vallis, valles	a valley on the surface of a planet or satellite.
vastitas	an extensive plain of lowlands on the surface of a planet or satellite.
Venerian	of or pertaining to Venus.
vent	the opening in the crust of a planet or satellite through which volcanic material erupts.
visual magnitude	*see* magnitude.
volume	the amount of space occupied by a three-dimensional object.
wind	the natural movement of atmospheric gases caused by differential forces due to the rotation of a planet or satellite.
y–d–h–m–s	short for years–days–hours–minutes–seconds; thus $2^y298^d16^h33^m20^s = 2$ years, 298 days, 16 hours, 33 minutes and 20 seconds.
year	generally defined as the time interval for the Earth or any planet to make one complete revolution around the Sun. Many other types of years are defined.
zenith	the point on the celestial sphere directly above the observer.
zodiac	a band around the celestial sphere 18° in width and centered on the ecliptic.

References and sources

Astronomical Tables of the Sun, Moon, and Planets, Jean Meeus, 1983, Willmann-Bell, Richmond, VA

The Biographical Dictionary of Scientists: Astronomers, David Abbott, 1984, Peter Bedrick Books, New York

Biography of Percival Lowell, A. Lawrence Lowell, 1935, Macmillan and Co., New York

The Cambridge Biographical Encyclopedia, David Crystal, 1995, Cambridge University Press

The Cambridge Guide to the Constellations, Michael E. Bakich, 1995, Cambridge University Press

A Chronicle of Pre-Telescopic Astronomy, Barry Hetherington, 1996, John Wiley & Sons, New York

The Dawn of Astronomy, Joseph Norman Lockyer, 1894, Macmillan, London

Early Astronomy, Hugh Thurston, 1993, Springer-Verlag, Berlin

Early Astronomy from Babylonia to Copernicus, William Matthew O'Neill, 1986, Sydney University Press, Sydney

From Vinland to Mars, Richard S. Lewis, 1976, New York Times Book Co., New York

A History of Astronomy, Antonin Pannekoek, 1969, Barnes & Noble, New York

The Immortal Fire Within – The Life and Work of Edward Emerson Barnard, William Sheehan, 1995, Cambridge University Press

Lowell and Mars, William Graves Hoyt, 1976, University of Arizona Press, Tucson, AZ

The Lowells and their Seven Worlds, Ferris Greenslet, 1946, Houghton Mifflin Co., Boston, MA

Mars, Percival Lowell, 1895, Houghton, Mifflin and Co., Boston & New York

Mathematical Astronomical Morsels, Jean Meeus, 1997, Willmann-Bell, Richmond, VA

Mercury, Faith Vilas, Clark R. Chapman and Mildred Shapley Matthews, 1988, University of Arizona Press, Tucson, AZ

Mysterious Universe: A Handbook of Astronomical Anomalies, William R. Corliss, 1979, The Sourcebook Project, Glen Arm

The Planet Neptune: an Exposition and History, John Pringle Nichol, 1846, William Tait, Edinburgh

The Planet Neptune – An Historical Survey before Voyager, Patrick Moore, 1996, Wiley-Praxis, Chichester & New York

Planets and Moons, William J. Kaufmann III, 1979, W. H. Freeman and Company, San Francisco, CA

Planets X and Pluto, William Graves Hoyt, 1980, University of Arizona Press, Tucson, AZ

Private research and public duty: G. B. Airy and the search for Neptune, Allan Chapman, *Journal for the History of Astronomy*, **19** (1988) 121–39

Satellites of Jupiter, David Morrison, 1982, University of Arizona Press, Tucson, AZ

Space Almanac, Anthony R. Curtis, 1992, Gulf Publishing Company, Houston, TX

The Timetables of Science, Alexander Hellemans and Bryan Bunch, 1988, Simon and Schuster, New York

Uranus – The Planet, Rings and Satellites, Ellis D. Miner, 1990, Ellis Horwood, New York

Venus Revealed, David Harry Grinspoon, 1997, Addison-Wesley, Reading, MA

Index

Note: General characteristics of each planet may be found (alphabetically) in the lists section and at the beginning of each of the individual planet chapters.